TURBINE TIP CLEARANCE MEASUREMENT

by

Lawrence C. Baker

Gordon E. Grady
and
Hagen R. Mauch

General Electric
Aircraft Engine Group

Propulsion Engineering Series

Wexford Press
2008

TABLE OF CONTENTS

LIST OF ILLUSTRATIONS

LIST OF ILLUSTRATIONS (Continued)

LIST OF ILLUSTRATIONS (Continued)

LIST OF ILLUSTRATIONS (Continued)

LIST OF ILLUSTRATIONS (Continued)

LIST OF TABLES

INTRODUCTION

BACKGROUND

Present turbine tip clearance design and control techniques are limited largely by the unavailability of accurate tip clearance measurement instrumentation. Without such instrumentation, data is not available to substantiate prediction correlations relating to thermal and centrifugal growth phenomena in turbine disks, blades, shrouds, and shroud support structures. This limitation results in the development of blade-shroud growth systems by trial and error and may yield larger than desirable turbine tip clearance gaps and/or a higher than necessary hardware usage rate. In addition, without tip clearance control, part power clearances may be larger than required and SFC is greater. All of these are costly from performance and procurement standpoints.

OBJECTIVES

The objectives of this program were to apply accurate turbine tip clearance measurement instrumentation to the full spectrum of actual gas generator operation and to determine the tip clearance response of typical rotor-shroud systems for future evaluation and improvement of design and control techniques.

RESULTS

This report presents a discussion of the design techniques used in establishing turbine tip clearances and describes the methods used in translating measured temperatures to clearance calculations. The results of the clearance calculations are compared to the measured and predicted clearances.

The program experience and test results with the laser tip clearance measurement device are discussed and evaluated, and recommendations for design improvements are presented.

TURBINE TIP CLEARANCE DESIGN CONSIDERATIONS

INTRODUCTION

Since clearance at the airfoil tips is a major source of component aero-dynamic performance loss, there is a need to minimize the average tip clearance of a turbine blade stage. Figure 1 shows the effect of clearances on turbine performance for a variety of engine sizes and configurations.

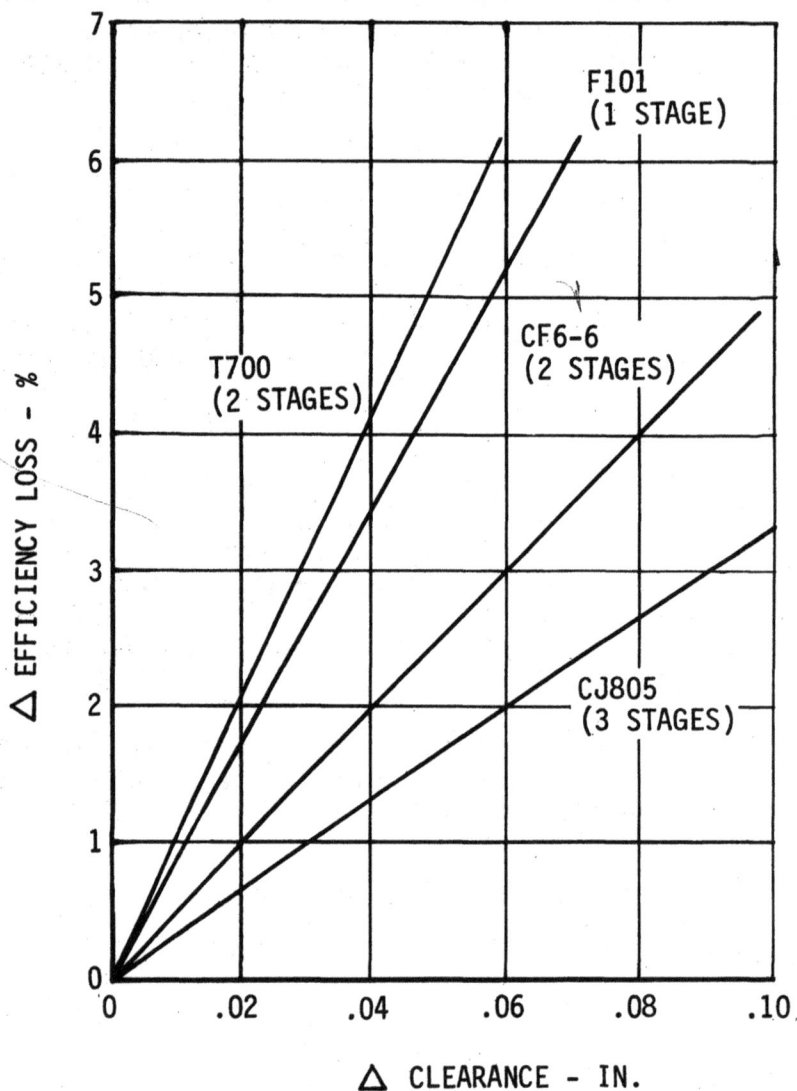

Figure 1. High-Pressure Turbine Efficiency Loss With Tip Clearances.

14

However, too tight a cold initial clearance causes excessive rubs that often progress to excessively large operating clearance, dangerous local overheating, freezing of rotors on shutdown, and can even cause engine operational failure. The balanced design provides tight average clearances at all operating conditions. To accomplish such a design, many conflicting considerations, estimates, calculations, and tests must be performed. The availability of a direct measurement clearance device would significantly aid in the establishment of tip clearance levels in the development phase of an engine program. In addition, such a device could be instrumental in defining turbine clearance control systems that would benefit engine performance. The following guidelines summarize the design approach for determining and maintaining tip clearance in turbines.

AERODYNAMIC CONSIDERATIONS

Turbine blade clearances are best compared on a nondimensional basis and on the basis of average operating clearance divided by blade length (C_L/L) for each stage of blades and vanes. The aerodynamic loss is estimated from this as $\Delta\eta_T$. Values for good C_L/L vary from 1.0 to 1.5% for a turbine.

TURBINE PERFORMANCE CONSIDERATIONS

It is important to note that as the net output of the thermodynamic cycle is decreased, such as at low power conditions and flight idle, the influence of component performance on cycle efficiency increases. Since operating speeds and temperatures also decrease at part power, turbine clearances may open, reducing component efficiencies. It is imperative to consider clearances at part power for those engines where part-power operation is significant in the flight envelope. Another consideration is the speed with which parts heat up or cool down. Specifically, a very heavy turbine rotor or a high radius-ratio compressor disk may not heat to steady state in less than five minutes after throttle advance to maximum, while the stator, casing and shroud supports may reach operating temperature in the first minute. At a constant measured turbine discharge temperature (T5), this means that the maximum thrust will build slowly up to the design value or will build up rapidly and then droop until steady-state clearances are attained. At constant required thrust, a higher than desirable turbine inlet temperature (T2) will result until steady-state clearances are achieved. Both conditions are undesirable, and can be minimized by a good clearance-controlled design.

CLEARANCE CONTROL PROCEDURE

Recognizing the importance of the effects of clearance on performance, the achievement of balanced "tight" operating clearance versus the physical need for "adequate" clearance at assembly must be obtained to avoid rubs during starting, throttle chops, rebursts, stalls, etc. A rigid formula for clearance design is not available; however, the following steps can be utilized:

1. In the early phases of design, the steady and transient temperatures of the rotor and stator shroud supports are calculated, as is the total rotor and stator deflections. From these, clearance changes are calculated during and after throttle burst, chop, and time delayed reburst. The methods for minimizing differences between rotor and stator radial growth by means of appropriate heating or cooling of each are considered. In past designs, controlling the stator growth by choice of cooling and material has been successful.

2. With consideration of the method of manufacture, assembly, and rotor bearing support locations, an estimate is made of variations of the rotor and stator diameters due to dimensional variation, eccentricity, flange looseness, effect of axial stackup on radial clearance of noncylindrical stages, effect of looseness of blades and vanes, casing out-of-roundness, and rotor and frame thrust deflection. Also considered are the effects of special operating requirements, such as maneuver loads.

3. The cold average and cold minimum clearances based on items 1 and 2 above are determined, as are the minimum and maximum stack clearances. The allowable clearances are set in a range somewhat tighter than the stackup prediction to allow for statistical variations. The rotor and shroud grind diameters are determined, including an allowance for engine rotor and stator eccentricities.

4. After careful comparison and evaluation of differences from the baseline design, blade lengths and rotor and stator diameters are specified. Blade and shroud grind diameters must also consider the flowpath contours. A typical small engine rotor and shroud are ground after assembly to a ± 0.001-inch tolerance. The shroud supports are first mounted in their casing, and a runout measurement is taken from the installed turbine rotor shaft and duplicates the rotor centerline. When the shrouds are ground, this runout measurement is duplicated in the fixture. This procedure insures that the rotor and stator (shroud) centerlines are concentric in the engine. In the design of the flanges locating the shroud, it is important that assembly fits be maintained to avoid shifts in alignment when the engine is in operation. When the mounting flange is cooled, it holds the shroud support through a rabbet which remains tight during engine operation. Buildup dimensions are 0.000 to 0.003 inch loose. Bolts through the flange are kept as tight as practical in their locating holes to prevent flange shifts.

5. During the development phase, all engine buildups and teardowns
    are monitored to check whether clearance requirements can be
    attained and whether rubs are excessive or nonexistent. As-
    built clearances are determined by direct measurement of tip
    diameters using "hard blades" (dovetail clearances removed by
    a clamp or wedge arrangement). Turbine blade tip loss versus
    initial assembled engine measured clearance is measured to help
    assess operating clearance. Normally, tip loss occurs on
    blades as a result of rubbing.

6. Based on the initial operational experience, clearances are
    modified to assure that occasionally only very light rubbing
    occurs on all buildups but not consistently all the time.
    Trends in rub location are investigated and causes of
    consistent rubs are eliminated. For example, if rubs appear
    in one area consistently, the introduction of eccentricity to
    grind diameters is considered.

7. Rotor and stator temperatures are measured early in the program
    to confirm calculated predictions of transient and steady-
    state clearance.

8. Installation problems are considered and those which cause
    stator hot and cold spots, casing or frame out-of-roundness,
    etc., must be avoided.

ROTOR AND STATOR RADIAL GROWTH MATCHING

To optimize growth of the rotor and stator during transients and steady-
state operation, consideration, such as the following, is given to the
problem of matching the thermal and elastic strain growths of the rotor
and stator.

1. Where part-power losses are not a primary consideration, the
    stator or shroud support structure is kept as cool as pos-
    sible, or low expansion alloy shroud supports are used. This
    will minimize the following rub-causing problems:

    a. Effect of rapid growth of rotor blade versus slow growth
        of stator during a throttle burst.

    b. Effect of relatively rapid shrinkage of stator and mis-
        match with slowly reacting rotor after a throttle chop
        due to thermal lag of the heavier rotor disk.

    In many applications, part-power performance is important.
    Here a match between rotor and shroud support or casing growth
    must be maintained over a wider area of operating speed and
    temperature. In most cases, the stator is cooled to match the
    rotor temperature. Beyond this, improvements may be obtained
    by use of low thermal expansion stators and shroud support.

17

2. Controlling rotor cooling with a known source and quantity of air results in more predictable rotor growth (and clearance). If disks are not cooled by forced convection, their temperature distribution due to free convection is difficult to predict. For this reason, free convection cooling is often avoided. If free convection heat transfer is unavoidable, disk cavity test data on similar designs should be consulted as a guide. Free convection heat transfer on a large disk results in slow temperature response to throttle bursts and chops and can cause poor rotor-to-stator match. Forced convection compressor discharge cooling air, which circulates on the forward and aft side of the disk faces and runs through the rim dovetail slots before entering the blade dovetails, is an effective means of rotor cooling. Additionally, inner balance piston leakage air to cool the disk ID is utilized. A rotor-forced convection cooling scheme is shown in Figure 2.

3. The calculation of stator growth considers the elastic strain of the stator due to internal pressure and, for internally unsupported nozzle vanes, the radial strains applied to the rails by radial nozzle reaction forces. The latter factor is negligible in small engines.

4. The temperature of turbine stators and shrouds can be controlled by compressor bleed air. Casings and shroud supports can be cooled by convection and impingement cooling air partly to avoid combustion hot-spot out-of-roundness in turbine shrouds. A very useful device in matching casing thermal response to rotor response has been the use of heating of the circumferential flanges to which shroud supports are usually bolted. By circulating small quantities of air through a flange (less than 1%), temperature response to throttle bursts is increased considerably and the flange steady-state temperature can be better controlled to the levels desired. During a throttle chop, the flange temperature will diminish slowly due to the reduction in cooling flow heat transfer. A cooled shroud support system is shown in Figure 2.

It is important to measure casing temperatures on each installation. This includes test cell and aircraft installations. Thermocouples are relatively easy to apply and a rapid accumulation of data is possible to check casing temperature uniformity in the installation.

ENGINE CASING BENDING DEFLECTIONS

Engine casings can and do bend due to engine thrust and maneuver loads. The nature of the bending deflection curve depends on the location of the thrust and other mount points with respect to the engine center line.

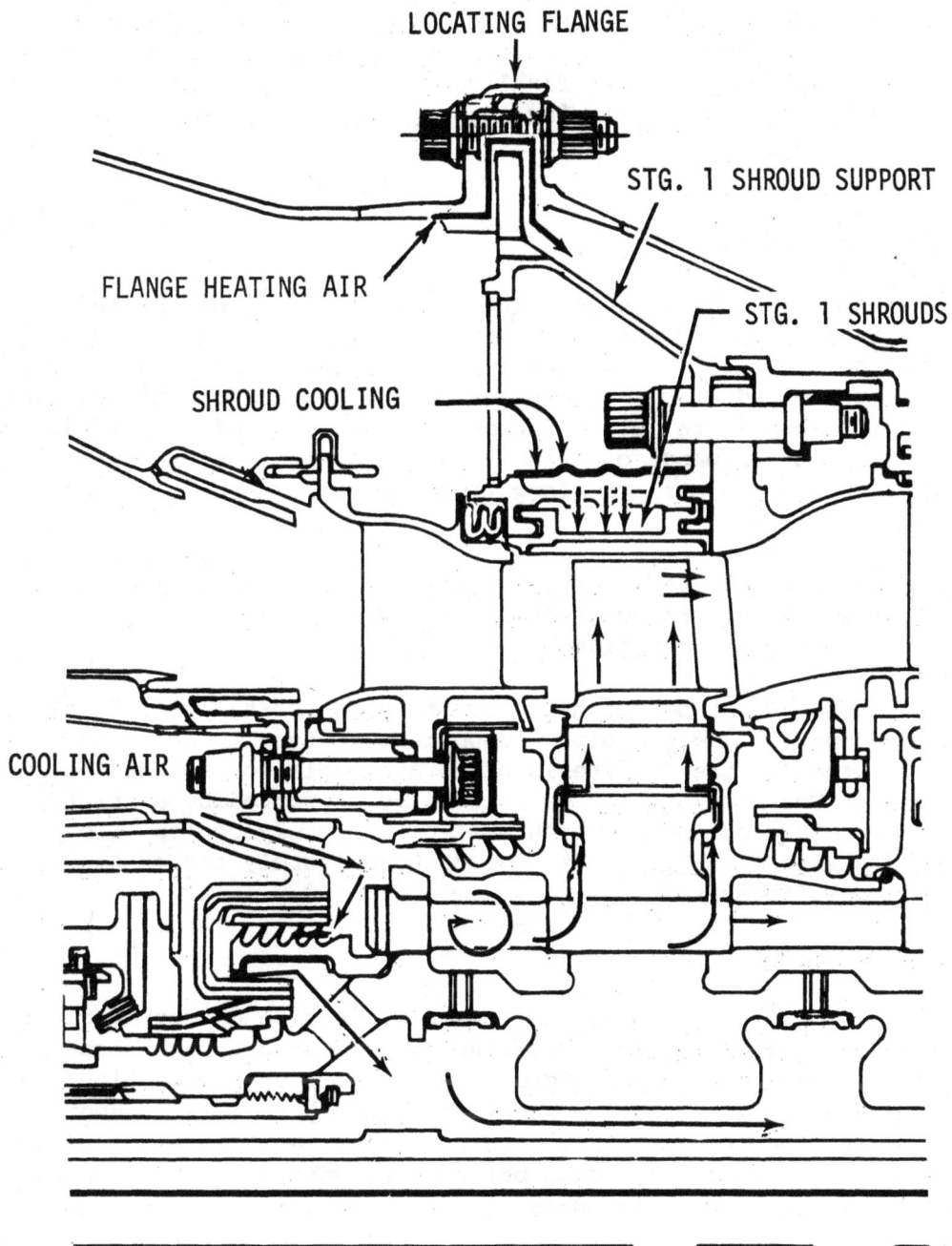

Figure 2. Stage 1 Disk and Stage 1 Shroud Cooling Systems.

Fan engines are especially subject to casing deflections due to engine thrust. The thrust mounts are often located far off the engine center-line, causing large shear and bending moments to exist which then deflect the engine casings. No serious problems exist if turbine rotors are straddle-mounted on frames. If rotors are overhung or located on "soft" oil-damped bearings, additional eccentricities between stator and rotor centerlines develop at zero and low speeds which can cause rub problems. The compressor and turbine designers take cognizance of engine mounting problems and provide inputs on the effect of mounting selection on per-formance prior to the point where final design decisions are made. Casing deflections can be estimated by using the appropriate analytical computer programs.

CASING ROUNDNESS CONTROL

Casings will remain round in service when supported by rigid bulkheads which do not permit significant nonuniform changes in radius. Two bulk-heads per casing are sufficient. One is not enough, since under nonuni-form externally applied forces on the casings, a single bulkhead can warp out of plane to accommodate out-of-roundness deflection of the casing. Two bulkheads cannot warp out of plane since this requires the cylindri-cal or conical casing to stretch along the meridian, where it is very stiff.

Strutted frames serve admirably as bulkheads, as long as there are a sufficient number of uniform temperature struts and stiff inner rings to maintain the roundness of the frames.

For turbine casings which do not have good frame support, it is possible to use turbine nozzles and/or shroud supports structurally as bulkheads. A frame at the rear of the turbine is often available to provide the addi-tional support. The nozzles must be internally supported by a stiff structure to form the bulkhead, which must be uniformly cooled to be effective. The shroud support must have deep, uniformly cooled sections to provide the stiffness required.

VANE AND SHROUD LOADS

Turbine shroud sector temperature and thermal expansion do not match the expansion of the shroud support. Nevertheless the shroud stiffness should not be allowed to influence radial growth so as to cause distortion of the shroud support. The most effective way to accomplish this is to segment the shrouds sufficiently to minimize their influence on radial deflection and shroud support distortion. Gaps between segments must be large enough to prevent "arch binding" during steady-state operation.

Turbine interstage nozzles are supported by the outer casing and often load the casing quite heavily, which affects turbine rotor and interstage seal clearance. It is important to note that the interstage seal ring stiffness of the nozzle sectors plays an important part in determining the reacting load on the casing. If nozzle sector arc lengths are long, the inner structure must be stiff in order to avoid lifting the nozzle from its radial supports. Short sectors or individual nozzles load the casing rails radially as well as tangentially and axially.

TURBINE SHROUDS AND SHROUD RUBS

High-temperature, high-performance turbine blades are unshrouded and operate with short shroud sectors, usually filled with a material such as Bradelloy. This material acts as an insulator and conducts little heat to the support. It allows its backing material to be cooled effectively by compressor air. At the high operating temperatures, the turbine tip rubs against the Bradelloy-filled shrouds, removing some of the blade material and depositing it on the shroud. This "scabbing" effect can result in extensive tip rubs and performance loss. To minimize this, it is best to run in a turbine so that rubs occur during a short time and at high speed. This will remove all the material from the blade tip that will ever be removed.

To avoid warping, shrouds should be short and well supported by circumferential rails with minimum clearance. At high temperature, most nickel base alloys and shrouds filled by Bradelloy oxidize and actually gain volume. Volume changes lead to dimensional changes and warping of the shrouds. To minimize this, the shrouds should also be effectively cooled, and a stiff backing structure should be included in the design. It is best to cast the shroud backing structure with ribs to stiffen it against warping.

21

CLEARANCE CALCULATION METHODOLOGY

SOURCE OF TEMPERATURE DATA AND TREATMENT

Temperature data used to calculate the transient clearances was obtained during tests of two YT700-GE-700 gas generators, Engine Serial No. 002-3A and Serial No. 008-5. Engine Serial No. 002-3A was instrumented with rotor and stator thermocouples measuring turbine metal and surrounding gas temperatures. Figure 3 shows the thermocouple locations. Data was obtained during five test runs at steady-state power settings of Idle, 39,000 rpm, 42,600 rpm, Intermediate Rated Power (IRP), and during transient starts, bursts, and chops. Data from the five tests were adjusted to a common inlet temperature for later use in rotor and stator temperature matching calculations.

Engine Serial No. 008-5 tests were run with thermocouples installed on the stator to confirm that temperature measurements from Engine Serial No. 002-3 could be used in clearance calculations for comparison with Engine Serial No. 008-5 measured clearances.

CALCULATION PROCEDURES

Temperature Distributions

For purposes of determining the detailed temperature distributions needed for calculation of turbine component growths, the following procedures were used. Finite element thermal models were constructed of both the rotor and the stator (Figures 4 and 5), which contained all features of the configuration contributing to positioning of high-pressure turbine rotor blade tip and stationary shroud. Heat transfer coefficients and gas temperatures on all surface boundaries were calculated on the basis of predicted and/or measured cycle gas conditions. Computer calculations were carried out using these initial models to determine predicted steady-state and transient temperature distributions of the turbine rotor and stator. Results obtained from this initial calculation were then compared with measured data to determine the accuracy of both initial and final calculated temperature values, and transient temperature response rates. Where necessary, boundary conditions were then altered to represent actual engine conditions more realistically, and computations were rerun until close agreement with measured temperature data was obtained.

Generally, calculated and measured temperatures were in close agreement for steady-state conditions. Data matching was applied primarily during transients to minimize the differences between measured temperatures and those used to compute clearances. In performing the data match calculations, care was taken not to include measured results which appeared questionable and which were not consistent with measurements in similar areas. Temperature measurements 1 and 15 at the rotor bore (Figure 6) represent an example where no explanation could be made for the low reading of thermocouple number 15 and its Idle temperature was, therefore, not matched.

22

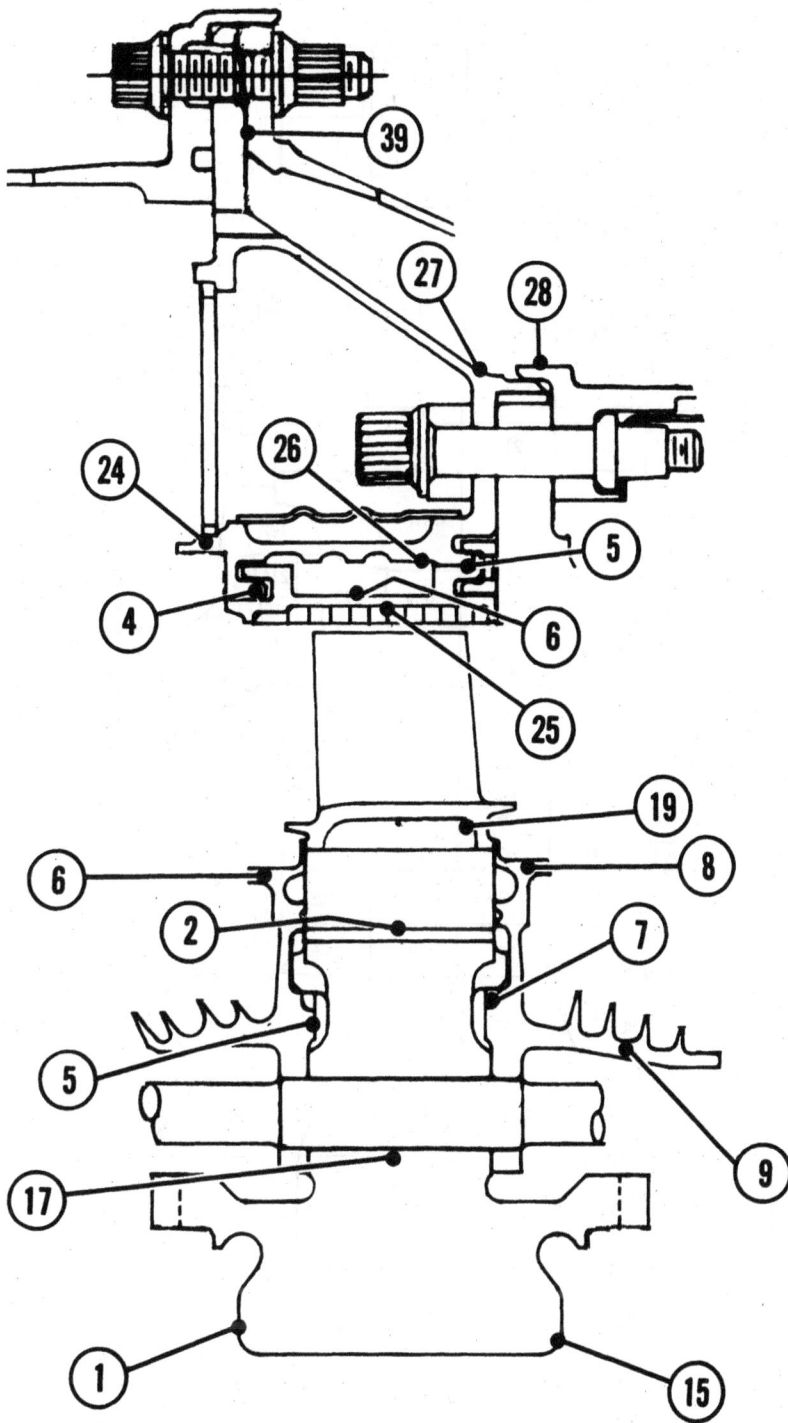

Figure 3. Location of Thermocouples on YT700 Engine Serial No. 002-3A.

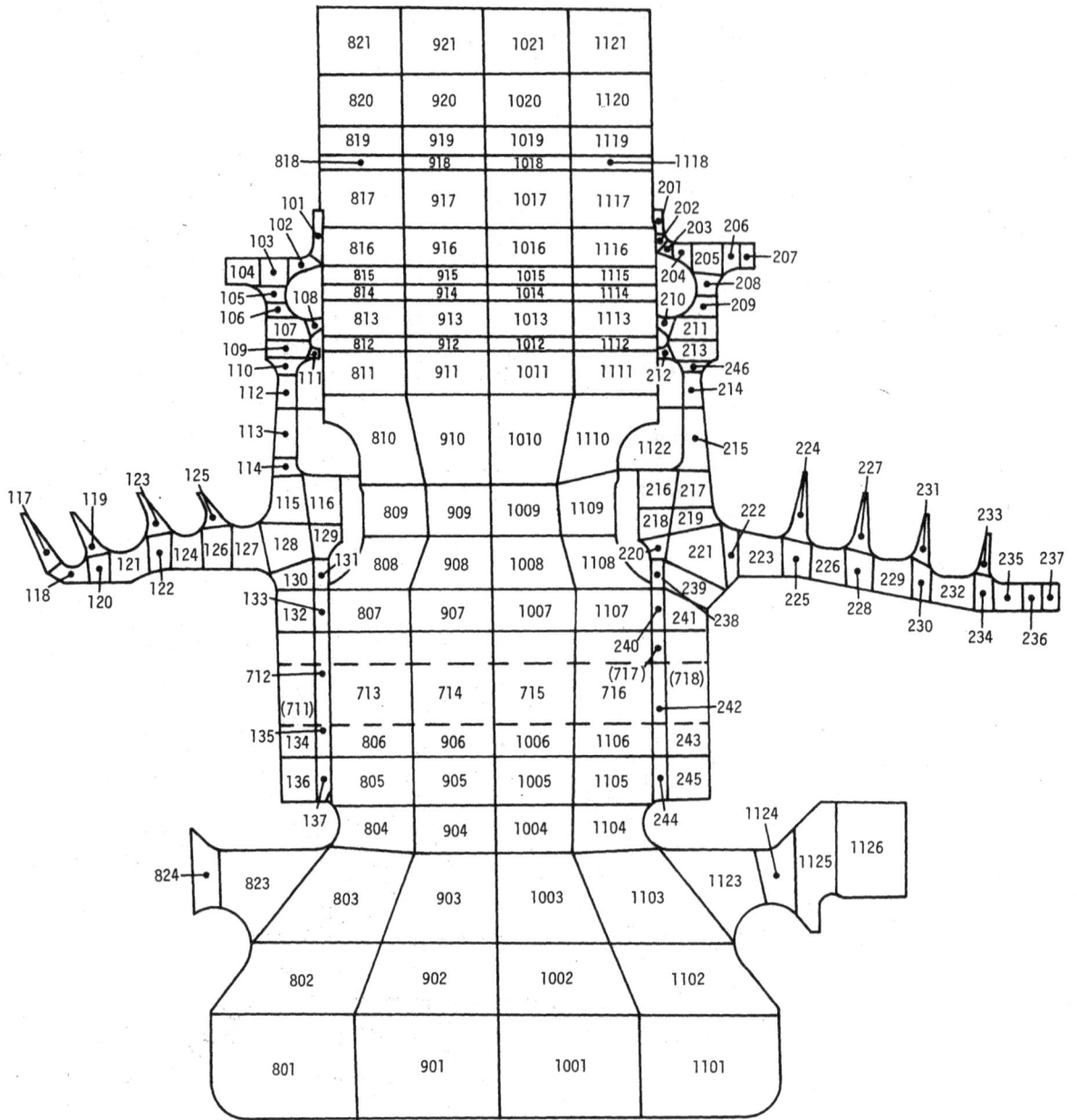

Figure 4. Stage 1 Turbine Rotor Nodal Model.

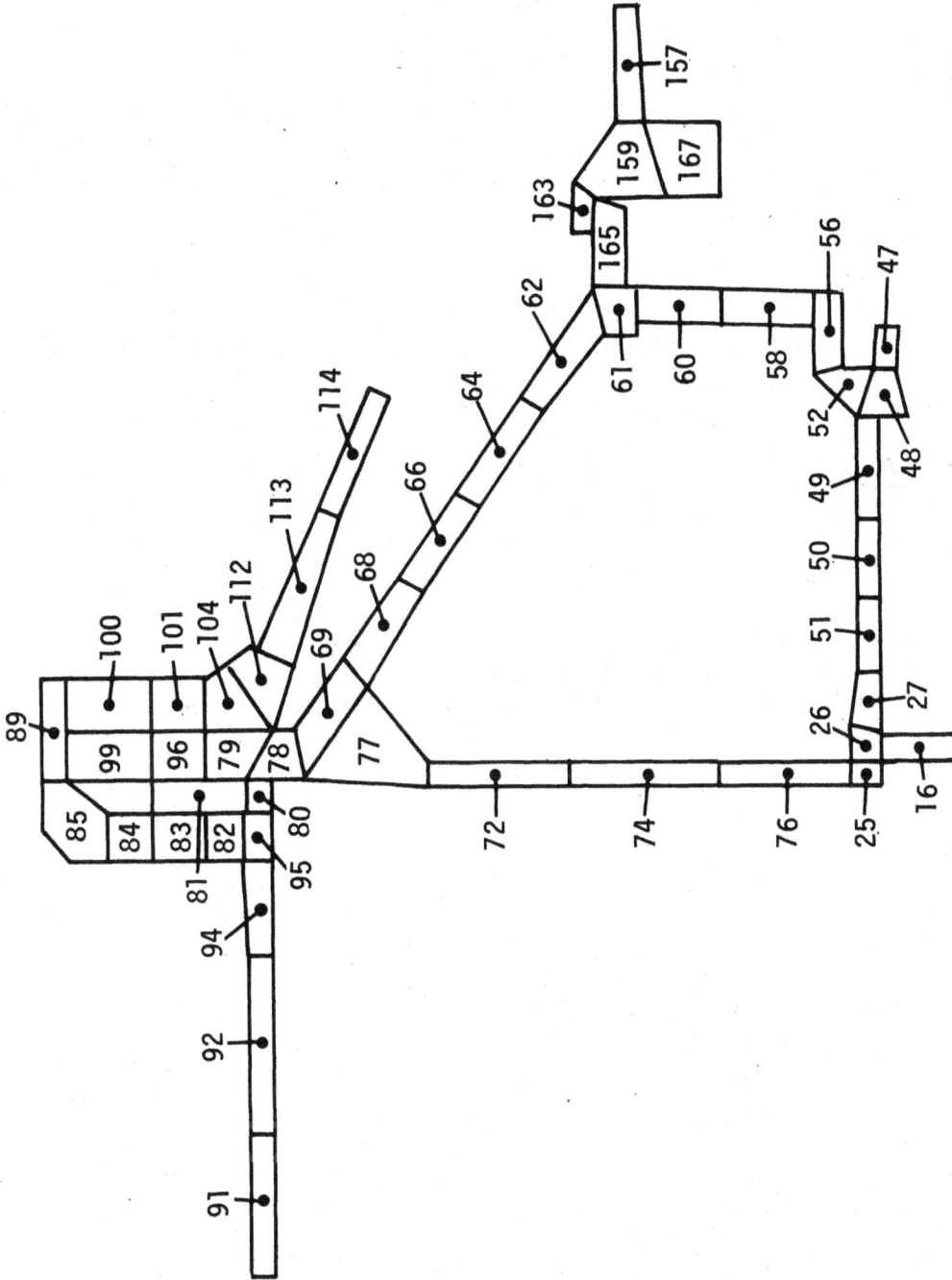

Figure 5. Stage 1 Shroud Support Nodal Model.

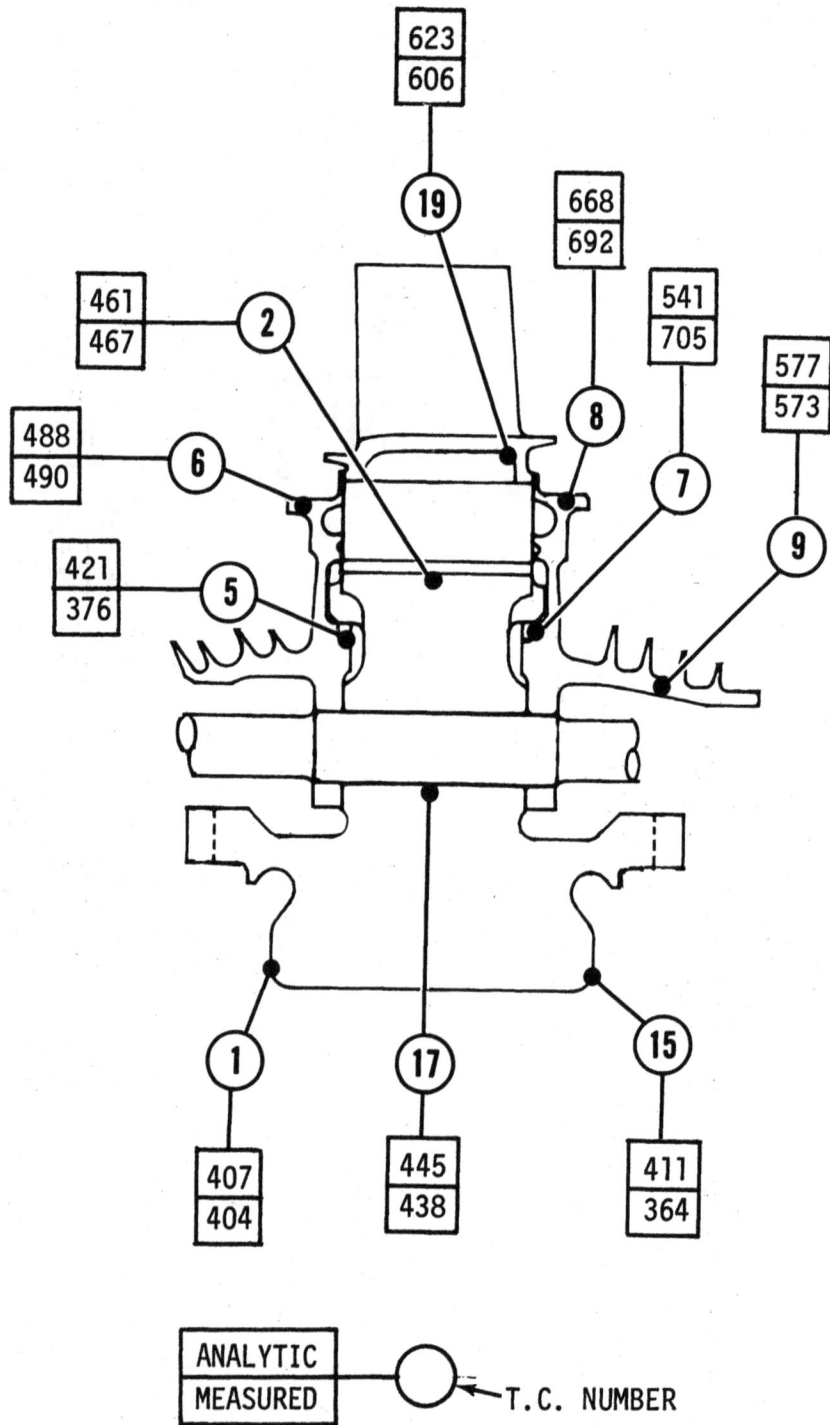

Figure 6. Stage 1 Turbine Rotor Idle Temperatures - Analytic vs Measured.

Once satisfactory agreement had been reached on the calculated and measured data matches, predictions were made for the start, burst chop, and reburst transient temperature distributions, and the results were applied for detailed growth calculations. Temperature response characteristics of each mode for each mission segment studied are shown in Appendix A.

Component Growth Calculations

In calculating clearances, three parameters are considered:

1. Turbine temperatures.

2. Centrifugal loads.

3. Casing and shroud support pressures.

To obtain the parameters, data is required which provides rotor and stator temperatures, rotor speed, and compressor discharge pressure as well as the relationship between compressor discharge pressure and casing and shroud support pressures.

Rotor Growth Analysis

Determination of the blade tip radial growth requires the summation of disk, disk dovetail, blade dovetail, and blade airfoil growths.

The mathematical model devised to analyze the disk growth is of the finite element type and is shown in Figure 7. As shown, this finite element model also incorporates the forward and aft cooling plates and the through-bolts, which are simulated as individual beams. Load boundary conditions at the forward and aft curvic coupling are based on prior design calculations which incorporated both stages of the T700-GE-700 high-pressure turbine rotor. Such a state-of-the-art model was previously applied at the steady-state design conditions only .

For the elements that are at a radius greater than that shown in the model of Figure 7, the blade and wheel dovetails and the blade airfoil, thermal and centrifugal effects have also been included in computing the growths.

Shroud Growth Analysis

The shroud growth is determined by the radial growth of the forward and aft shroud supports. For this analysis, the finite element model shown in Figure 8 was applied. This model simulates the holes and slots that are incorporated in the design. Boundary conditions at the model cutoff sections were scaled from prior design calculations which incorporated both stages of the YT700-GE-700 high-pressure turbine stator and shroud support. The present model is more detailed than one previously used for the shroud support studies and provides improved calculation accuracy.

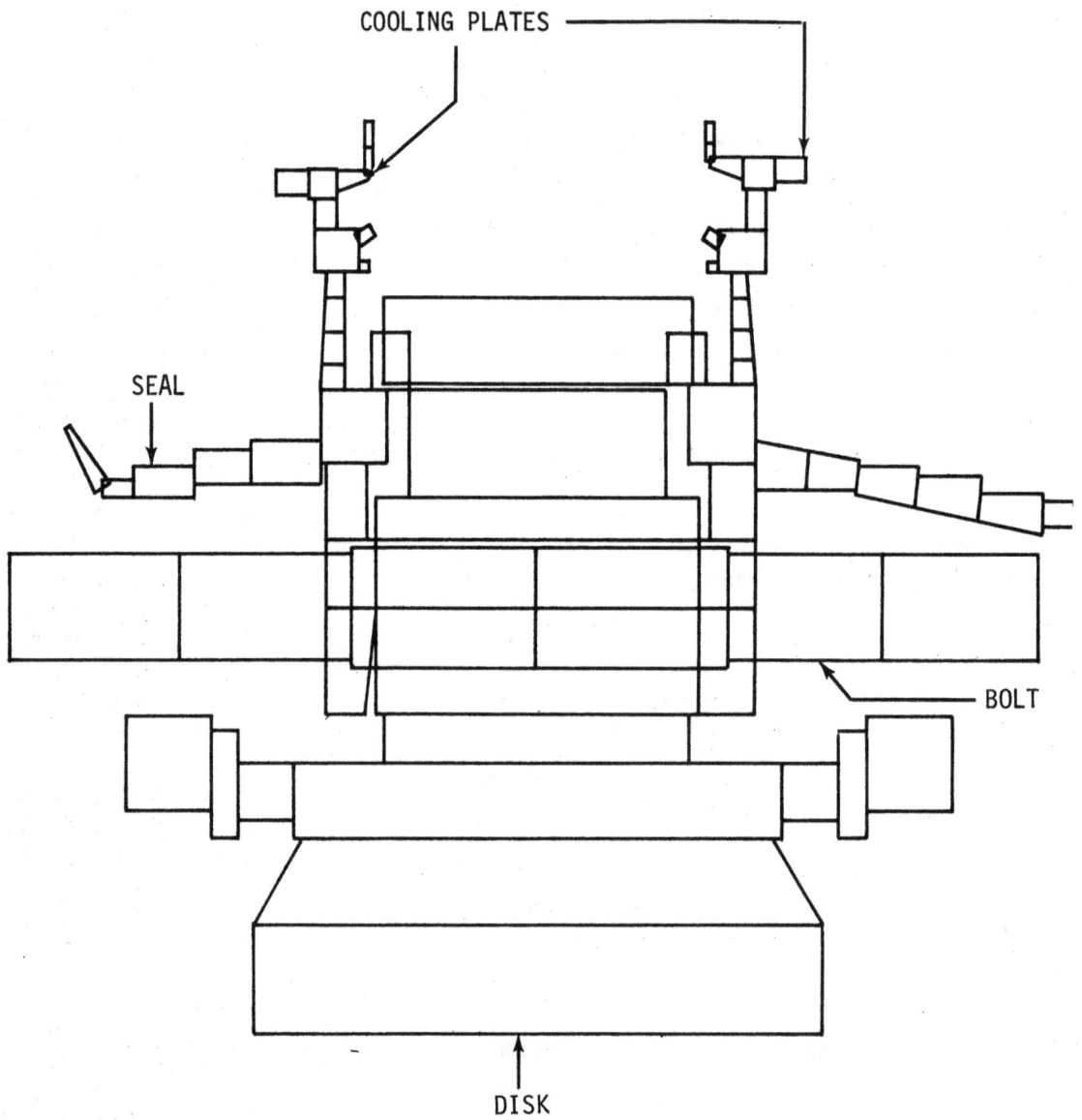

COOLING PLATES

SEAL

BOLT

DISK

Figure 7.   Stage 1 High-Pressure Turbine Rotor Finite Element Model.

28

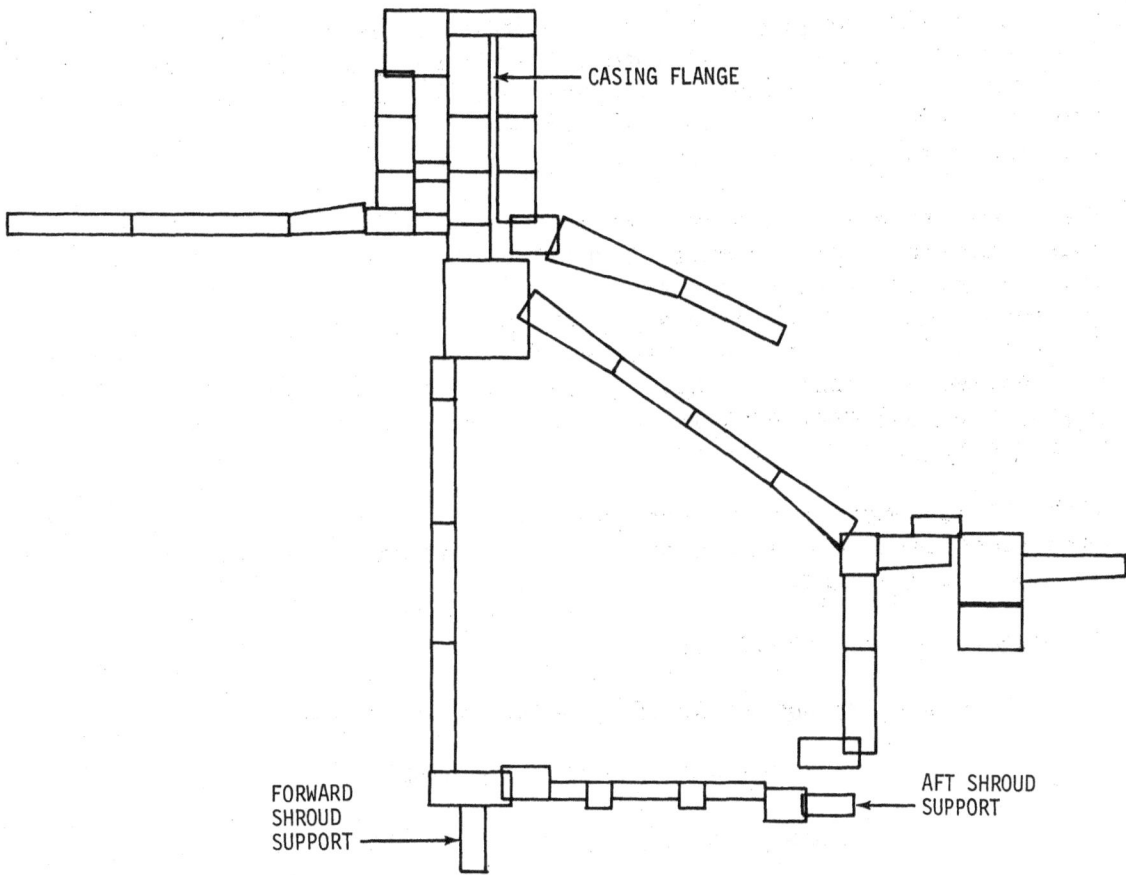

Figure 8.   Stage 1 High-Pressure Turbine Stator Finite Element Model.

Pressures in the high-pressure turbine shroud support area required for the analysis were obtained by correlating with compressor discharge pressure (P_{s3}) using the relationship shown in Figure 9. The data shown were measured on T700-GE-700 Engine Serial No 014-3A.

During an engine transient, a small lag or overshoot of cavity pressures versus P_{s3} is anticipated. Experience from other engines indicates that such lags or overshoots occur only within the first one or two seconds of a transient and have little effect on stator growth. Thus, the pressure relationships shown in Figure 9 were used to evaluate cavity pressures for transients as well as steady-state conditions.

Clearance Calculations

The finite element model contains the procedures for calculating growth due to thermal, centrifugal, and pressure loading. The hardware is modelled by shells, rings, and beams with each joint identified by a number, a nominal radius, and a thickness. This information is summarized in the geometric computer input file.

The temperature matrix previously described assigns a temperature to each joint, providing the computer thermal input file. Using this temperature input file, the thermal expansion coefficient and modulus of elasticity for each element is determined by the program. The centrifugal load is determined from the nominal radius, rotor speed and mass of the element between two successive joints. Loads due to pressure are found by multiplying the projected area of an element between two successive joints with the respective pressure.

With thermal expansion and complete loading established, all boundary conditions between elements are satisfied by continued stress and displacement iterations.

Tip clearances are found as:

    Clearance   = Engine Buildup Measured Clearance

                 + Average Shroud Support Growth

                 - Disk Growth - Dovetail Growth - Blade Growth

30

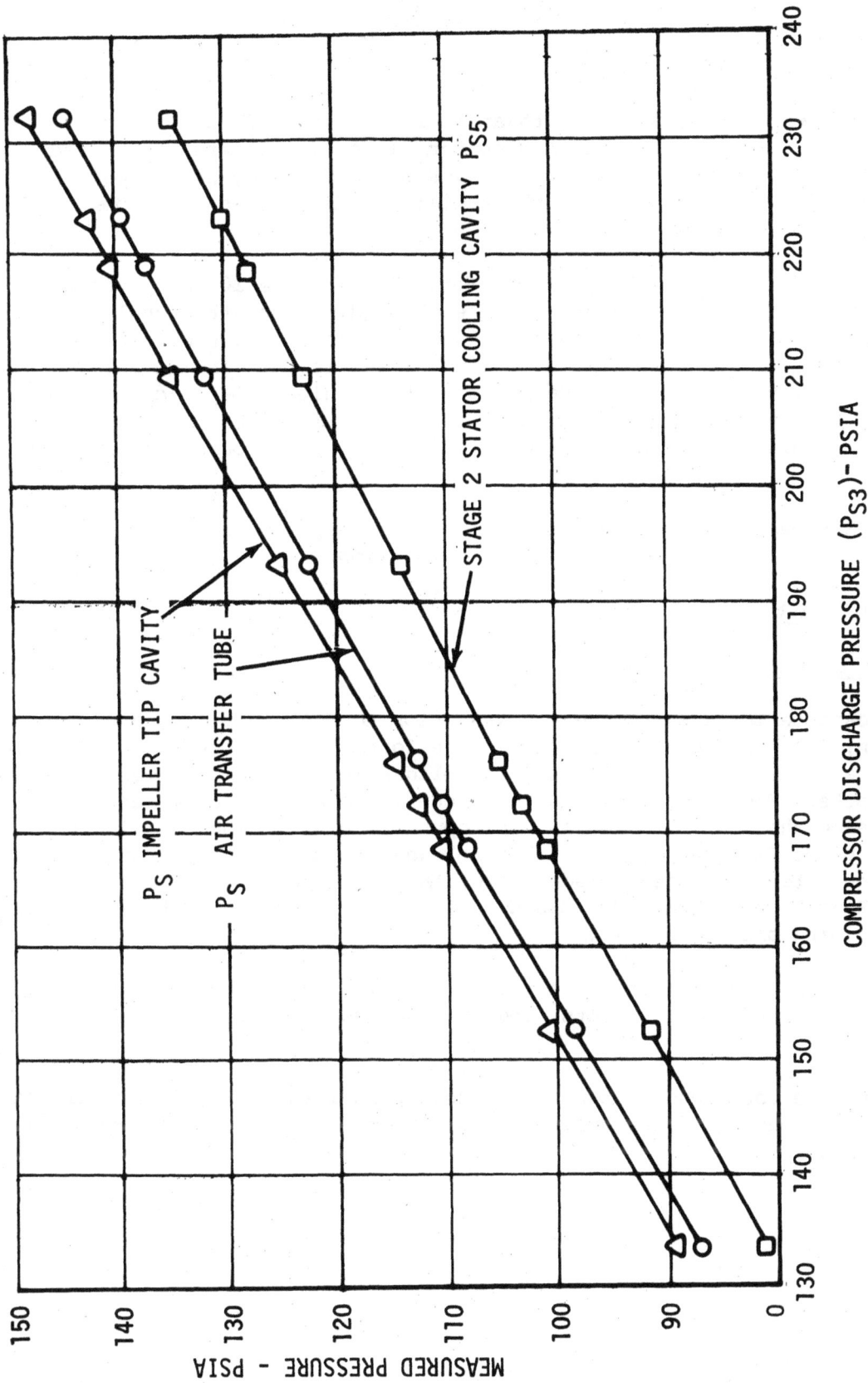

Figure 9. Relationship of Compressor Discharge Pressure to Static Cavity Pressure.

31

CLEARANCE ANALYSIS

CALCULATED CLEARANCES

The results of the hardware growth analysis using the data matched temperatures from YT700-GE-700 Engine Serial No. 002-3A are summarized in Tables 1 through 5 for the selected transient and steady-state conditions. Figure 10 shows the location of the radial points at which the component growth history is presented.

In Figure 11, total rotor tip growth (Item 3 in Figure 10) and average shroud growth (Item 6 in Figure 10) are plotted. Since this analysis is independent of the measured engine cold clearance, a value of 0.014 inch was assumed for this data (the 0.014-inch value is the minimum allowable for YT700-GE-700). Thus, Figure 11 shows the minimum clearances of the YT700 configuration for the conditions analyzed with the minimum clearance 0.014 inch occurring about 10 seconds into a burst from idle to IRP.

Figure 12 shows radial closure for the burst, chop, and hot reburst. Prior predictions have been added for comparison.

MEASURED CLEARANCES

Engine Cold Clearances

Cold clearance measurements were taken prior to engine buildup and after teardown; no significant differences were found. The procedure for taking these measurements is as follows. The blades were shimmed outward and set on V-blocks. The longest blades were identified and the diameter determined using the longest blade and the opposite blade. The radial runout of each blade (longest blade equal zero) was taken. With the runout of blade opposite the longest blade known, the radius of the maximum length blade (Rmax) was determined. Based on the runouts and Rmax, the radii for the remaining blades are known.

The shroud radii were found in a similar manner. The minimum diameter was obtained, and using runouts, the average and minimum radii were established.

Combining the rotor and stator measurements resulted in the determination that the minimum cold clearance was 0.0155 inch and the average equalled 0.0302 inch.

32

| Time (sec) | Radial Growth [a] (mils) | | | | | | Net Closure 3 - 6 |
| | Rotor | | | Stator | | | |
	Location 1	Location 2	Location 3	Location 4	Location 5	Location 6	
0	0	0	0	0	0	0	0
15	.9	1.2	4.3	0	0	0	+4.3
17	1.2	1.6	6.5	0	.1	.1	+6.4
20	2.1	2.6	7.5	.1	.6	.4	+7.1
25	2.3	3.0	7.9	.5	1.5	1.0	+6.9
30	2.5	3.5	8.4	1.0	2.4	1.7	+6.7
35	2.8	4.1	9.0	1.5	3.4	2.3	+6.5
45	3.4	5.1	10.0	2.4	5.2	3.8	+6.2
65	4.5	6.7	11.6	3.8	8.0	5.9	+5.7
85	5.3	7.7	12.6	5.0	10.1	7.6	+5.0
105	5.9	8.4	13.3	6.1	11.7	8.9	+4.4
125	6.4	9.0	13.9	7.0	12.8	9.9	+4.0
145	6.8	9.5	14.4	7.8	13.5	10.7	+3.7
165	7.1	9.8	14.7	8.5	16.1	11.3	+3.4

TABLE 1.   ANALYSIS OF A START TO IDLE

(a)   See Figure 10 for radial location definition

TABLE 2. ANALYSIS OF A BURST FROM IDLE TO INTERMEDIATE RATED POWER

| Time (sec) | Radial Growth[a] (mils) | | | | | | Net Closure 3 - 6 |
| | Rotor | | | Stator | | | |
	Location 1	Location 2	Location 3	Location 4	Location 5	Location 6	
0	8.2	10.9	16.1	10.9	15.4	13.2	+2.9
2	4.2	12.7	19.1	11.1	15.8	13.4	+5.7
5	10.5	14.3	22.8	11.8	17.0	14.4	+8.4
10	11.6	15.9	26.7	13.2	19.3	16.3	+10.4
15	12.2	16.9	27.8	14.3	21.2	17.8	+10.0
20	13.1	18.0	28.9	15.4	22.7	19.1	+9.8
30	14.3	19.6	30.5	17.0	24.7	20.9	+9.6
50	16.5	22.1	33.0	19.3	26.9	23.1	+9.9
70	17.0	22.7	33.6	20.9	28.2	24.6	+9.0
90	17.7	23.5	34.4	21.9	28.9	25.4	+9.2
110	18.1	24.0	34.9	22.5	29.3	25.9	+9.0
130	18.4	24.3	35.2	22.8	29.6	26.2	+9.0
150	18.6	24.5	35.4	23.1	29.8	26.5	+8.9

(a) See Figure 10 for radial location definition

| Time (sec) | Radial Growth [a] (mils) | | | | | | Net Closure 3 - 6 |
| | Rotor | | | Stator | | | |
	Location 1	Location 2	Location 3	Location 4	Location 5	Location 6	
0	19.1	25.0	35.9	23.5	30.0	26.8	+9.1
2	18.3	23.9	32.2	23.4	29.9	26.7	+5.5
5	16.3	21.6	26.8	23.2	29.5	26.4	+0.4
8	16.2	21.2	26.4	22.9	28.9	25.9	+0.4
12	15.9	20.6	25.8	22.5	28.1	25.3	+0.5
17	15.5	19.8	25.0	21.9	27.1	24.5	+0.5
25	14.9	18.8	24.0	21.1	25.6	23.4	+0.6
35	14.1	17.7	22.9	20.1	24.1	22.1	+0.8
50	13.1	16.5	21.7	18.9	22.2	20.6	+1.1
70	11.9	15.0	20.2	17.5	20.3	18.9	+1.3
100	10.8	13.7	18.9	15.7	18.3	17.0	+1.9
125	10.1	12.9	18.1	14.5	17.3	15.9	+2.2
150	9.6	12.3	17.5	13.5	16.7	15.1	+2.4
200	9.0	11.7	16.9	12.2	16.0	14.1	+2.8

TABLE 3.   ANALYSIS OF A CHOP FROM INTERMEDIATE RATED POWER TO IDLE

(a)   See Figure 10 for radial location definition

| Time (sec) | Radial Growth[a] (mils) | | | | | | Net Closure 3 - 6 |
| | Rotor | | | Stator | | | |
	Location 1	Location 2	Location 3	Location 4	Location 5	Location 6	
0	15.3	18.0	23.2	21.6	26.5	24.1	−0.9
2	16.3	19.8	26.2	21.7	26.6	24.2	+2.0
5	17.6	21.4	29.9	21.9	26.9	24.4	+5.5
10	18.1	22.4	33.2	22.2	27.5	24.9	+8.3
15	18.2	22.9	33.8	22.4	28.0	25.2	+8.6
20	18.4	23.3	34.2	22.6	28.3	25.5	+8.7
30	18.5	23.8	34.7	22.9	28.8	25.9	+8.8
50	18.7	24.3	35.2	23.2	29.3	26.3	+8.9
70	18.8	24.5	35.4	23.3	29.6	26.5	+8.9
90	18.9	24.7	35.6	23.4	29.8	26.6	+9.0
110	19.0	24.9	35.8	23.4	29.9	26.7	+9.1
130	19.0	24.9	35.8	23.4	29.9	26.7	+9.1
150	19.0	24.9	35.8	23.4	30.0	26.7	+9.1

TABLE 4.   ANALYSIS OF A HOT REBURST (20 SECONDS AFTER CHOP)

(a)   See Figure 10 for radial location definition

| Condi-tion | Radial Growth[a] (mils) | | | | | | Net Closure 3 - 6 |
| | Rotor | | | Stator | | | |
	Location 1	Location 2	Location 3	Location 4	Location 5	Location 6	
IDLE	8.2	10.9	16.1	10.9	15.4	13.2	+2.9
Ng= 39,400	13.8	17.6	24.1	17.6	22.7	20.2	+3.9
Ng= 42,575	16.9	20.5	30.0	21.2	27.3	24.3	+5.5
IRP	19.1	25.0	35.9	23.5	30.0	26.8	+9.1

TABLE 5.   ANALYSIS OF STEADY-STATE CONDITIONS

(a)   See Figure 10 for radial location definition

Touch Probe Measurements

Touch probe measurements were made during Engine Test No. 3 on YT700-GE-700 Engine Serial No. 008. The touch probe is basically a steady-state device which measures minimum clearance (see Touch Probe Design Description, Turbine Tip Clearance Measurement Device Section). During this test, by rapidly retracting and re-inserting the probe manually, some measurements were also obtained during transients.

The touch probe zero calibration was performed as shown in Figure 13. The process showed good repeatability and resulted in a value of Δ = 0.243 inch (arbitrary reference value). Clearances measured by the touch probe are found using the following relationship:

Touch Probe Clearance = (Blade Touch Reading - Flag Reading) - Δ

Data was obtained at Idle, during a 1-minute accel to IRP, a 1-minute decel to Idle, a burst from Idle to IRP and at steady-state conditions as shown in Figure 14. Prior predictions and the refined calculations documented in this report have also been plotted in Figure 14 to show a direct comparison.

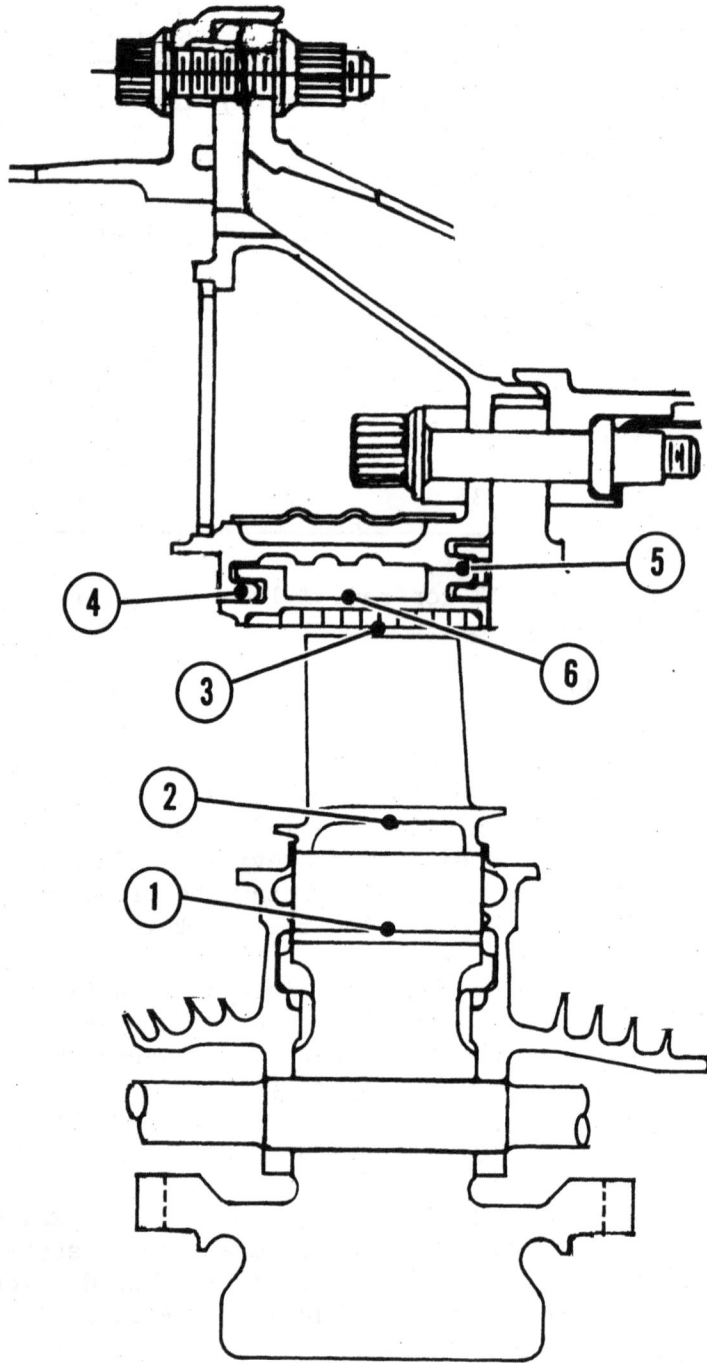

Figure 10. Radial Growth Calculation Points.

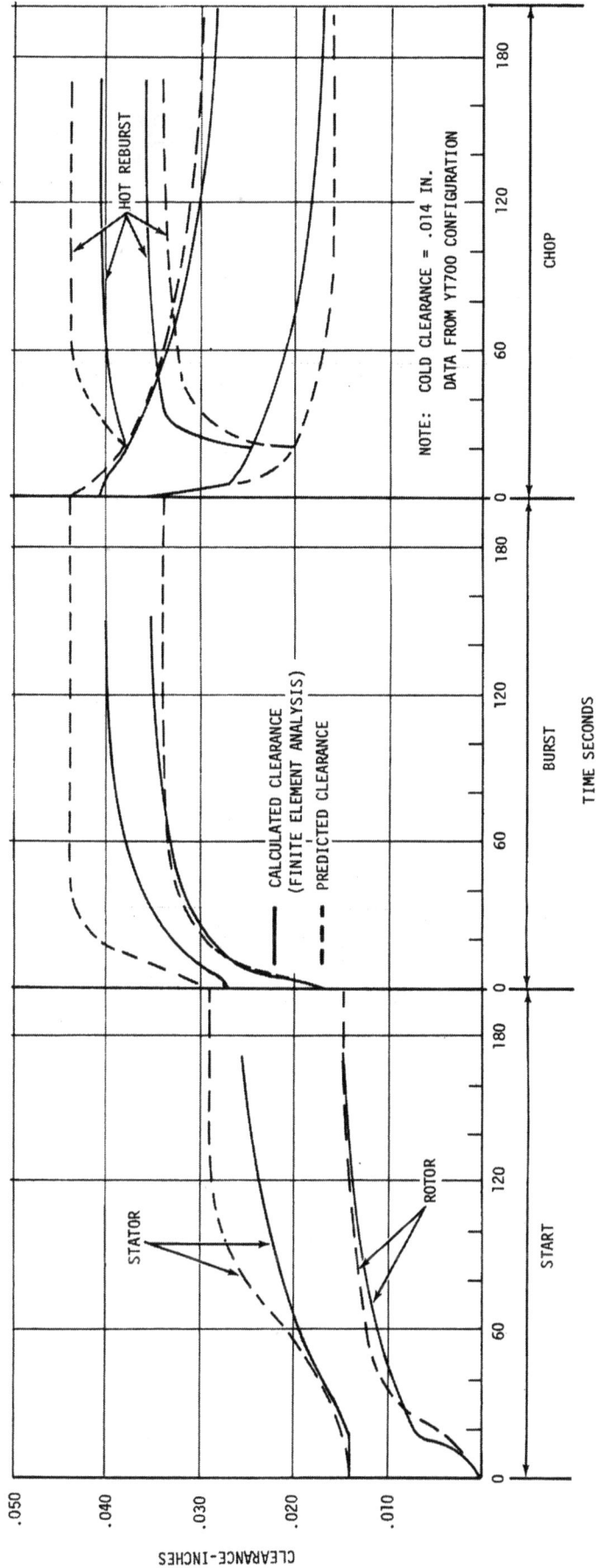

START   BURST   CHOP

TIME SECONDS

CLEARANCE-INCHES

HOT REBURST

STATOR

ROTOR

CALCULATED CLEARANCE
(FINITE ELEMENT ANALYSIS)

PREDICTED CLEARANCE

NOTE: COLD CLEARANCE = .014 IN.
DATA FROM YT700 CONFIGURATION

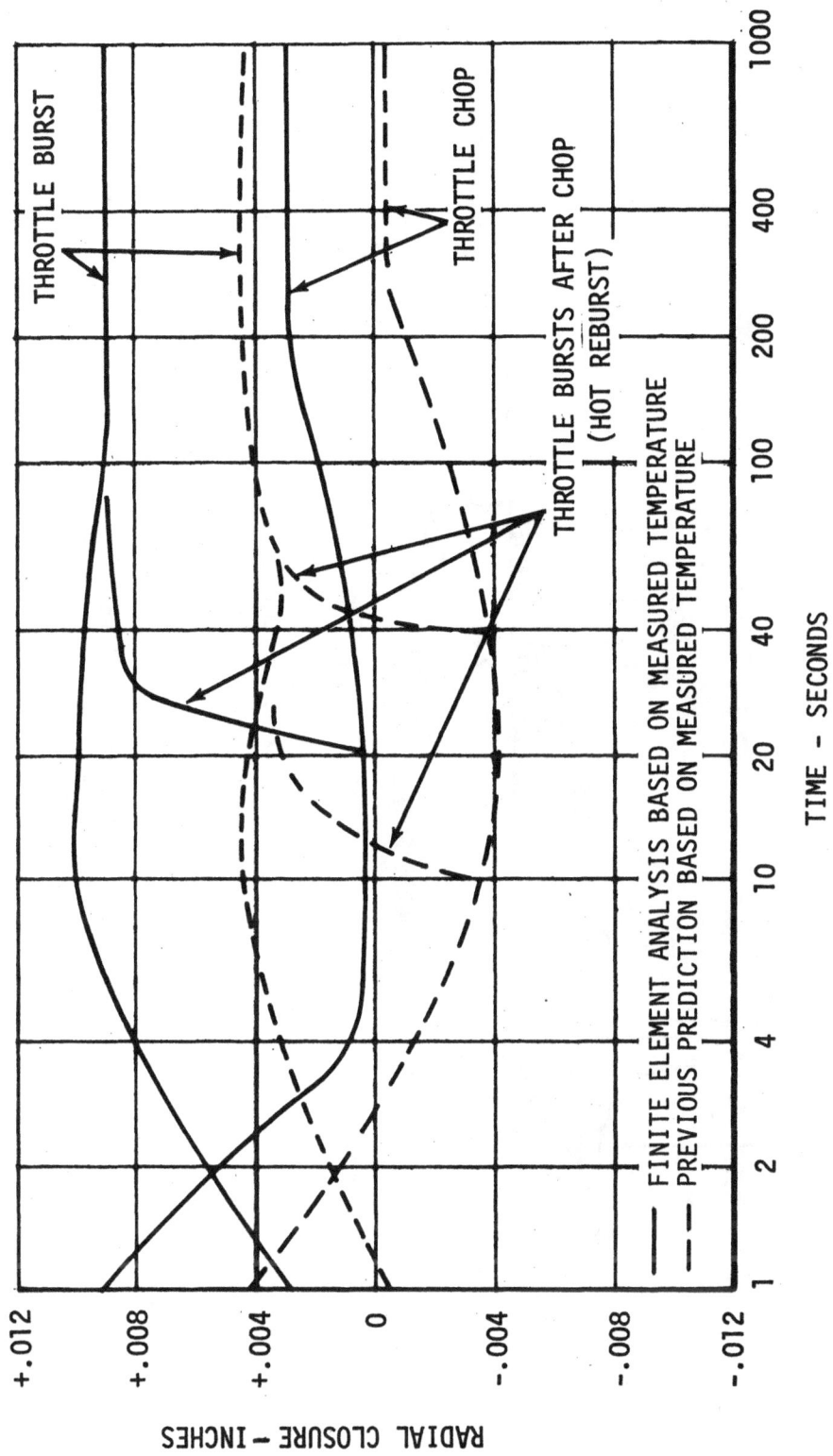

Figure 12. Calculated and Predicted Radial Closure.

41

FLAG IN RETRACTED POSITION

CALIBRATION DEVICE
(FLUSH WITH SHROUD I.D.)

PROBE GUIDE

STEP 2

FLAG IN MEASURED POSITION

PROBE GUIDE

STEP 1

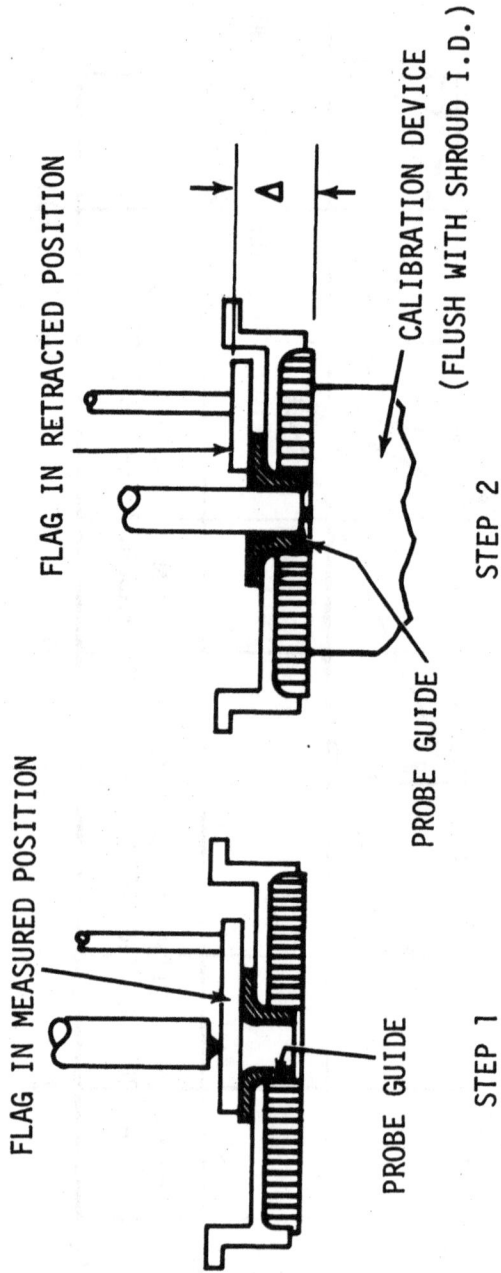

FIND ZERO CALIBRATION Δ AS FOLLOWS:

TAKE STEP 1 AND STEP 2 READINGS AND
SUBSTRACT RESULT.

REPEAT PROCESS TO PROVE
REPEATABILITY

Figure 13. Schematic of Touch Probe Zero Calibration.

42

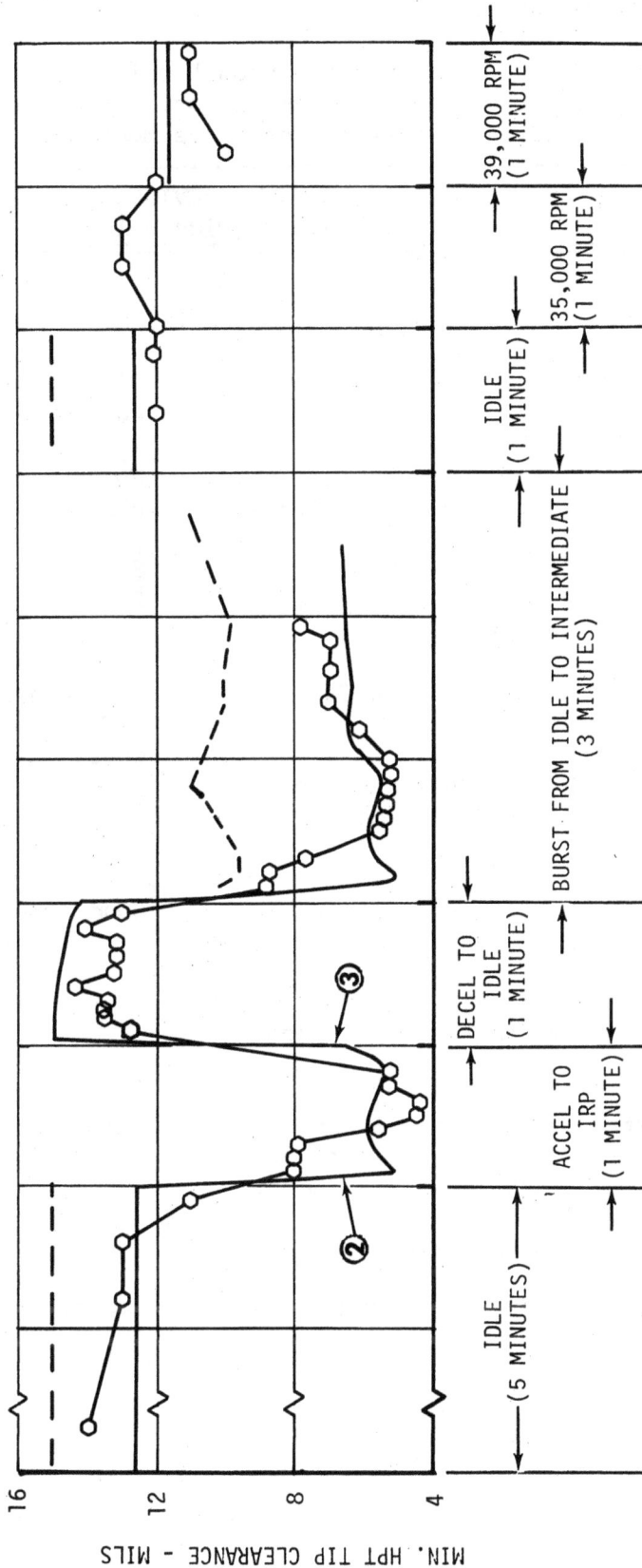

Figure 14. Comparison of Measured, Predicted, and Calculated Clearances.

43

DISCUSSION OF RESULTS

The clearances resulting from the growth calculations, shown in Figure 11, reveal significant differences between prior predictions and the refined predictions. The major differences are summarized in Table 6.

TRANSIENT	ROTOR	STATOR
Start	Reasonable agreement.	Significant difference after 40 sec.
	Initial growth is faster than previous analysis.	Up to 5 mils difference at around 2 minutes.
	Long term growth is slower than previous analysis.	Stator continues growth beyond 200 seconds.
Burst	Good agreement for first minute.	Significant difference; particularly around 30 sec where up to 8 mils difference occurs.
	Rotor continues growth beyond 200 seconds.	
Chop	Significant difference, for the first 5 seconds.	Good agreement, particularly in the first minute.
	Good agreement for time greater than 200 seconds.	
Hot Reburst	Significant difference. However major part of disagreement is due to initial growth difference during the chop.	Significant difference, except at T = O.

TABLE 6.   COMPARISON BETWEEN PREVIOUS AND REFINED ANALYSIS
(Reference Figure 14)

The radial closure, shown in Figure 12, indicates that the character of the closure curves versus time is in reasonable agreement for both analyses, but for all conditions the closure predicted by the refined technique is larger. Related to engine operation this would indicate that operating clearances are smaller than previously predicted.

The maximum closure of 10.4 mils is predicted to occur at 10 seconds into a burst from idle to IRP. During a start, the maximum closure is 7.1 mils. Thus, a typical mechanical and transient engine break-in procedure will decrease tip clearances by approximately 3 mils beyond the minimum encountered during a start.

Applying the calculated closure to Engine Serial No. 008-5 post test minimum clearance of 15.5 mils allows a direct comparison with the touch probe measurements. The results plotted in Figure 14 indicate good agreement of the refined calculations (within approximately one mil) in most instances. Differences between touch probe readings and calculations are expected when rapid transients are being made since the touch probe requires five seconds or more to be cycled from one reading to the next. Accuracy of the touch probe may also be affected by forcing rapid readings.

Since the refined calculations and the earlier predictions were based on the same temperature data (Engine Serial No. 002-3A) and since the touch probe data and the refined calculations are in good agreement, it would appear that the refined calculations represent engine clearance conditions more accurately.

As shown in Tables 1 through 5, the aft flange (location 5) of the YT700-GE-700 shroud support configuration grows more than the forward support (Location 4), resulting in a tilt of the shroud during engine operation. The T700-GE-700 qualification test (QT) engine configuration substitutes INCO 903 material for INCO 706 material in the rear part of the shroud support. Since the thermal expansion coefficient of INCO 903 is approximately 50% of that of INCO 706, the QT engine shroud support should experience little or no tilting.

TURBINE TIP CLEARANCE MEASUREMENT DEVICE

INTRODUCTION

The turbine tip clearance measurements required on this program were to be taken with the miniaturized tip clearance measurement device (MTCMD)[1] designed at Pratt and Whitney Aircraft (P&WA), Florida Research and Development Center. Since this design utilized a video display, which is basically unsuited to real time electronic condition monitoring and control, General Electric proposed the substitution of a solid-state position detector with the capability of direct analog electronic readout for display, recording, or further on-line analysis.

Initial laboratory tests verified the P&WA bench test data. However, testing and analysis showed that the system would have to be modified in order to deliver good electronic data from a high-speed, high-temperature turbine rotor.

Modifications which could be accomplished within the scope of the program were made to the MTCMD, but these changes were not suffficient to overcome the severe operational and environmental demands of engine testing. One principal obstacle to the successful demonstration of the system was the loss of a large percentage of the light energy in the system, some of which was detrimentally scattered within the probe (noise). Aggravating this situation was the variable and generally small amount of optical energy reflected from hot, rapidly rotating blades. As a result, the clearance signal impinging on the position detector was insufficient to properly trigger that device.

This section describes the systems; details the tests and the difficulties encountered and the solutions applied; and presents recommendations for improving the design of this type of device.

DESIGN DESCRIPTION

System Specifications and Criteria Summary

The MTCMD was originally designed, built, and tested by P&WA. General Electric proposed to utilize this basic concept with changes in the detector and readout areas. The GE system uses a position detector and electronic readout in place of a TV system.

---

1. Ford, M.J., Hildebrand, J.R., and Prosser, J.C., DESIGN, FABRICATION, AND DEMONSTRATION OF A MINIATURIZED TIP CLEARANCE MEASURING DEVICE, Pratt and Whitney Aircraft, USAAMRDL Technical Report 74-67, U.S. Army Air Mobility Research and Development Laboratory, Fort Eustis, Va 23604, September 1974, AD 787318

In order for a turbine tip clearance measurement device to be a viable tool, it was established that it meet the design criteria summarized in Table 7 (see Reference 1).

TABLE 7. MINIATURIZED TIP CLEARANCE MEASUREMENT DEVICE DESIGN CRITERIA			
REQUIREMENT	PARAMETER	MTCMD PART	DESIGN CRITERIA
Mechanical	Available Space	Probe Body	
Environmental	Temperature	Probe Tip	1900°F
		Probe Body	1000°F
		Probe Head	500°F
		Laser and Television Camera	150°F
	Pressure	Probe Body	10 Atmospheres
	Vibration	Probe Assembly	±1G from 50-250 Hz
			±2.5Gs from 250-500 Hz
			±10Gs from 500 Hz-1K Hz
			±20Gs from 1-2.5 K Hz
Operational	Range	System	0 - 0.065 in.
	Accuracy	System	±0.001 in.
	Sensitivity	System	±0.0005 in.
	Response Rate	System	0.4×10^{-6} sec
	Cooling Flow Rate	Probe Assembly	0.016 lb/sec or less
	Power Consumption	System	280 Watts or less
Cost		System	Low
Life		System	50 hr or greater

Miniaturized Tip Clearance Measurement Device System

The MTCMD is an optical distance gauge using the light beam triangulation method. A narrow beam of light emerges from the tip of the probe at a fixed angle. Longitudinal displacement of the target causes a transverse displacement of the return beam. This, in turn, is imaged on the face of a coherent (image-transmitting) fiber optic bundle at the rear of the probe. Thus, displacement of the target is converted to the motion of a light spot across the output fiber optic bundle.

The MTCMD system consists of a probe containing optical components, a laser light illumination system, a television camera to detect the position of the output spot, and data display and recording instruments. The overall system is illustrated in Figure 15.

Theory of Operation

Laser: The light source is a helium-neon laser. A standard microscope objective focuses the beam to a narrow spot less than the size of a single optical fiber. The fiber optic illumination system allows the laser to be mounted away from the engine casing. Given the requirement of illuminating no more than a single fiber, a laser is an ideal light source. The almost perfectly collimated beam is easily focused to microscopic size. Another advantage of a laser is its spectral purity, which allows detection of its radiation through extrememly narrow band optical filters, and the consequent rejection of infrared and visible lights which are not at its wavelength.

Probe: The probe contains the essential optical components for blade tip triangulation, and mounts in the casing of the first-stage turbine. Light is admitted to the probe through a fiber optic bundle in which a single (0.001 inch) fiber is illuminated. This has the advantage of an extremely small source size; however, the characteristic radiation pattern of an optical fiber is a cone of light. Since the input fiber optic bundle is mounted off the probe axis, the cone of light fills the spherical surface of a plano-convex lens and converges to an imperfect focus at the opposite side of the probe. A prism then directs the beam towards the target at a fixed angle. The beam reflects off the target and is imaged on the face of the output fiber optic bundle via the same prism and lens (Figure 16). The output of the MTCMD probe is a light spot whose position on the output fiber optic bundle is indicative of blade tip position. In practice, each blade will be under the probe only a few microseconds. Therefore, the output of an MTCMD system is a series of flashing light spots, each at a position corresponding to the clearance of its respective blade tip.

Detector and Readout: The P&WA MTCMD system uses a vidicon camera tube to detect the position of the reflected light spot on the output bundle. The camera uses a standard TV lens with an accessory "close-up" lens in order to form an enlarged image of the output bundle through the narrow band optical filter.

Output from the camera is fed through a cable out of the test cell to a television monitor and video tape recorder.

48

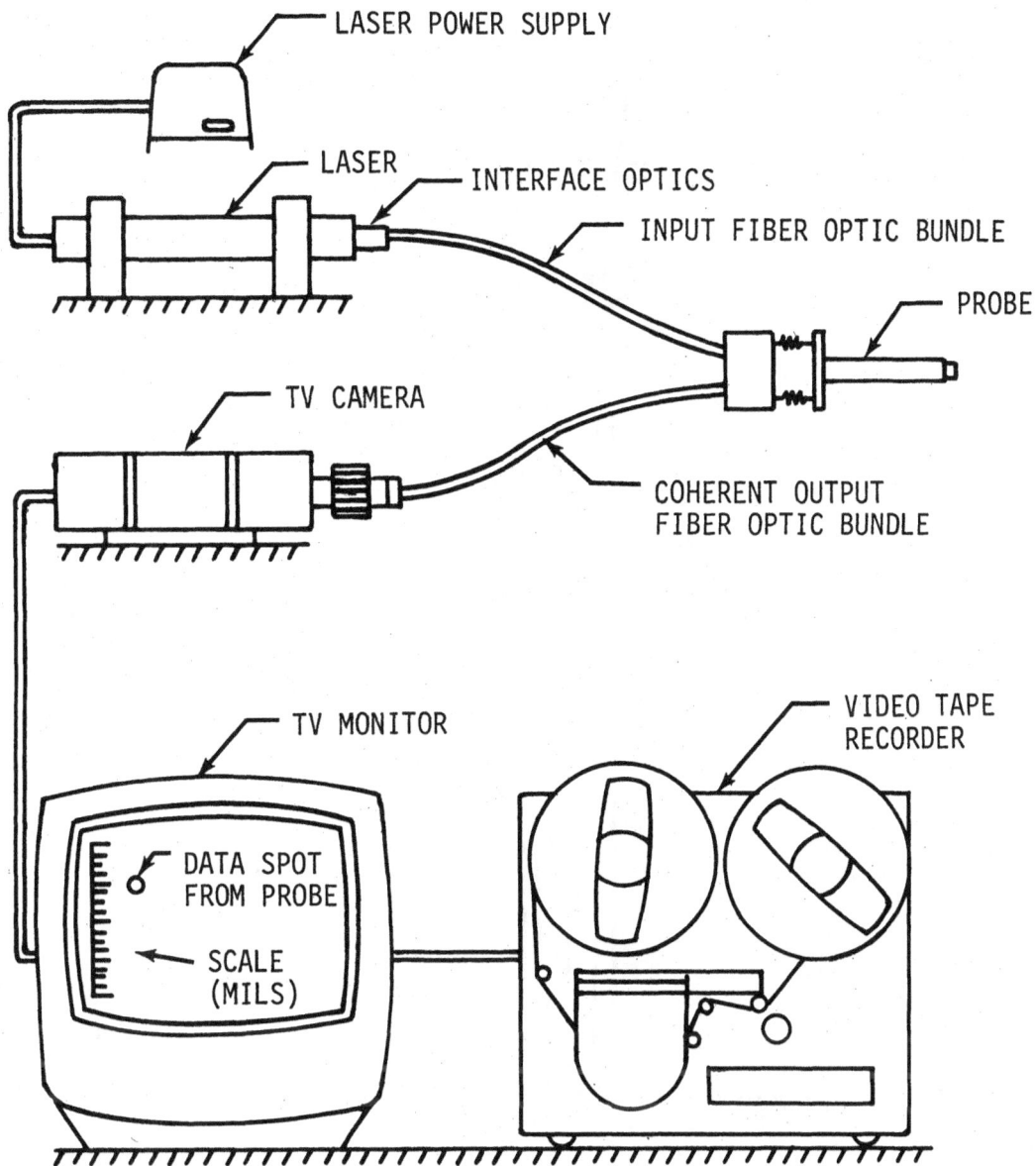

LASER POWER SUPPLY

LASER   INTERFACE OPTICS

INPUT FIBER OPTIC BUNDLE

PROBE

TV CAMERA

COHERENT OUTPUT
FIBER OPTIC BUNDLE

TV MONITOR

VIDEO TAPE
RECORDER

DATA SPOT
FROM PROBE

SCALE
(MILS)

Figure 15.   MTCMD - System Configuration.

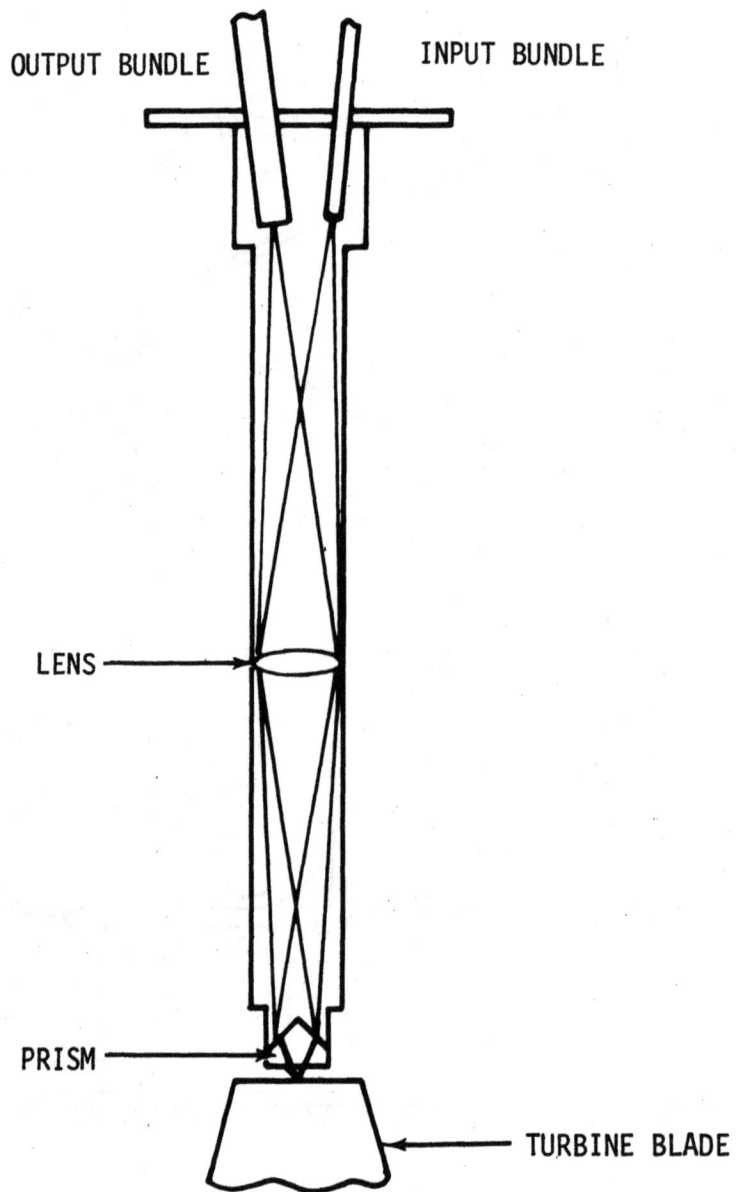

OUTPUT BUNDLE

INPUT BUNDLE

LENS

PRISM

TURBINE BLADE

Figure 16. MTCMD - Probe Configuration.

50

The TV camera makes a series of 30 full screen pictures per second (525 horizontal scans per picture) of the output bundle, detecting pulses as they occur during the scans. The TV monitor has a phosphor which retains each image for a length of time. Therefore, if a majority of blades are at or near one clearance value, they will integrate to a bright image on the screen at the corresponding position. A single long or short blade will not be defined on the TV monitor due to sampling rate and the time-averaging characteristics of the phosphor.

In general, a tight distribution of clearances provides a small spot on the screen. A large, but even distribution of clearances provides an elongated spot. An uneven distribution is weighted towards the clearance indications with the greatest frequency of occurrence, histogram style.

The MTCMD system, therefore, responds basically to the time-averaged value of tip clearance.

General Electric System

The GE system utilizes the same type of laser and probe as the P&WA system. The major difference is in the detector and readout system whose description and theory of operation follow.

This system uses a solid-state position detector with sufficiently fast response to give a clearance indication for each blade as it passes the probe tip. The system block diagram is shown in Figure 17.

The detector chosen for this application is an SC-10 continuous position sensing diode. The SC-10 is a Schotty barrier photo diode with a uniform resistive film which conducts the photo current from the incident light spot to two end terminals. Any deviation of the centroid of the light spot from the center of the device results in a current imbalance in the two terminals proportional to the displacement from the null position (center) and the incident power.

The SC-10 diode has a square junction surface of approximately 0.4 inch on a side. In this application, only one of the two perpendicular axes is used. Rated sensitivity is approximately one microamp per milliwatt per 0.001 inch displacement (at the Helium-Neon laser wavelength of 632.8 nanometers) and rated response time is 6 microseconds. The current output of the SC-10 diode is treated electronically through a signal conditioning circuit to provide a voltage output proportional to the tip clearance.

The signal conditioning circuit consists of a current-to-voltage amplifier, a voltage sum and difference circuit, and a high-speed divider (shown in Figure 18).

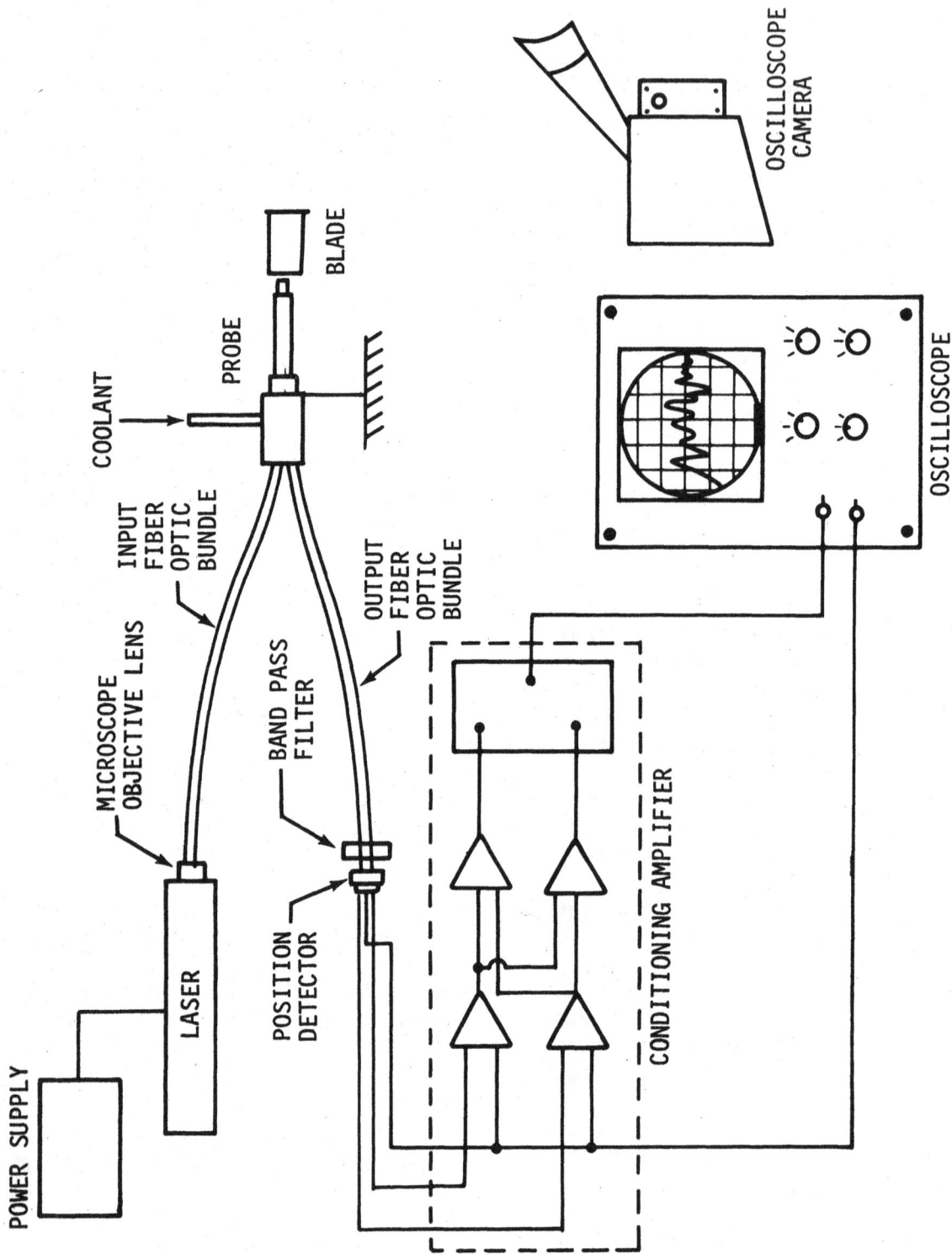

Figure 17. General Electric Tip Clearance Measurement System

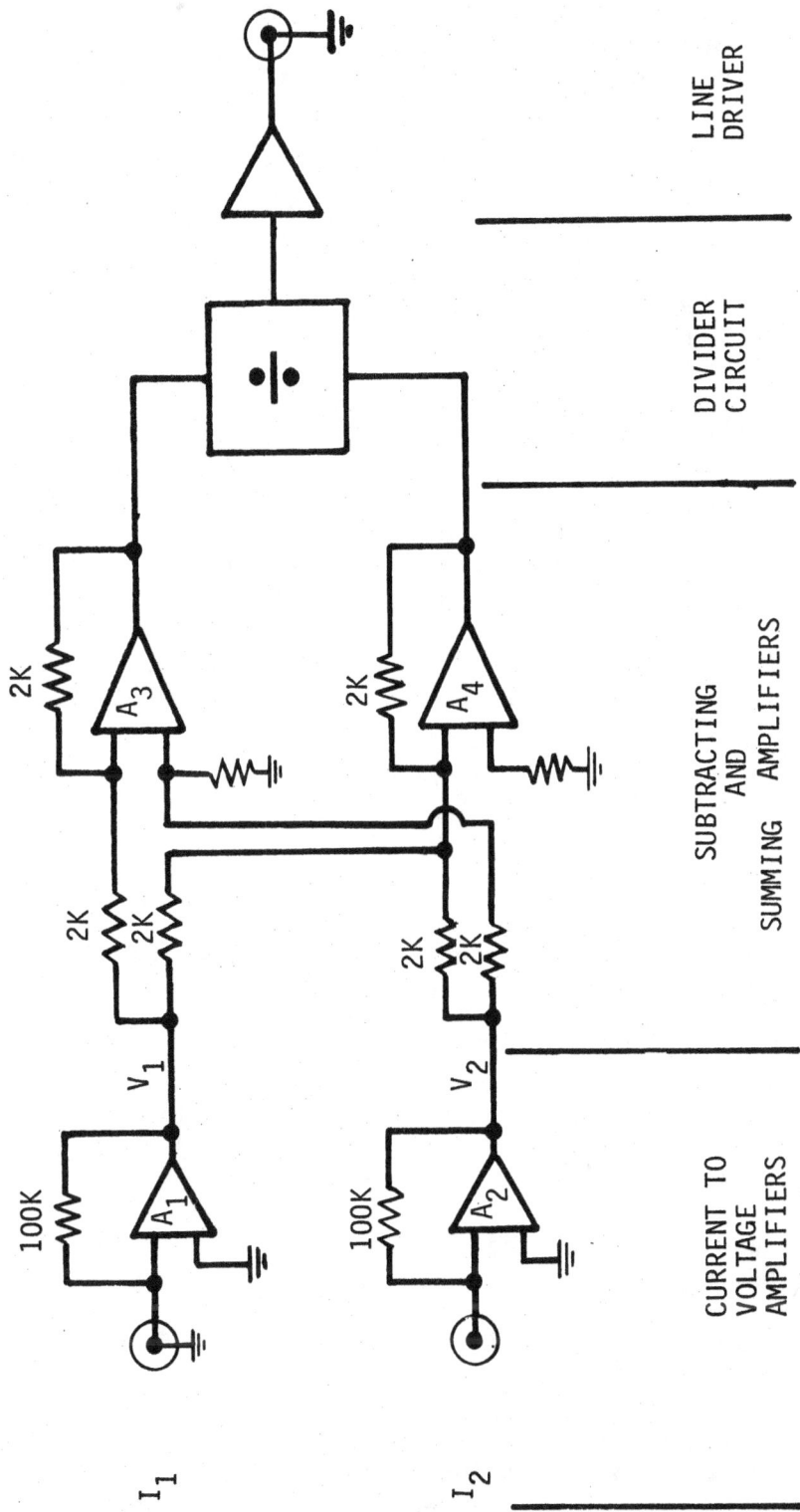

Figure 18. Signal Processing Circuitry.

Current from each diode leg is converted to usable voltages in the first two high gain amplifiers A1 and A2. Amplifiers A3 and A4 form the difference and sum of these respective voltages. An analog divider module takes the ratio of these two signals. A line driver feeds the clearance indication to CRT and recording equipment.

Touch Probe

General Electric provided an alternate clearance measurement device which could be used as an additional check on the accuracy of the MTCMD. The touch probe is designed to give a reading of the distance of the longest blade to the casing (minimum blade clearance).

A touch probe consists of a wire surrounded by coaxial layers of a magnesium oxide insulator and a protective outer shell. The probe is precisely driven until it contacts a reference surface (a moveable spring-loaded "flag" in the probe passage). A reading is then taken of this distance. Then the reference flag is turned aside and the probe is driven through a hole in the casing and shroud until it contacts the highest blade tip. A momentary electrical contact causes the control mechanism to stop the probe. A reading is taken, and the probe is withdrawn for another measurement. The difference of the two readings (minus the thickness of the flag plus casing and shroud) gives the actual minimum clearance. Figure 19 shows the touch probe system.

PRELIMINARY LABORATORY TESTS

Test Results

Early work consisted of bench testing various components of the MTCMD. A cutaway probe housing was used to facilitate observation. The following results were obtained:

1. The optimum angle of reflectance was determined to be 19 to 38 degrees depending upon target to prism distance. This result was in agreement with the P&WA report and their designed hardware.

2. It was determined that the laser beam could be reflected from a hot target.

3. The loss from the fiber optic bundle was found to be 11 percent per foot.

4. The resolution of the MTCMD probe, with a highly reflective nichrome target, was determined to be 0.0005 inch. A calibration of the "shiny" target is shown in Figure 20. For a 50-mil range of clearance the output varied from 0.11 volt to 6.48 volts. The curve is nonlinear where the nonlinearity is a function of the electronics.

Figure 19. Touch Probe System.

10 MILLIWATT LASER, CUT-AWAY PROBE

POLISHED TARGET (NICHROME) AT ROOM TEMPERATURE
USING THE LINEAR DIODE

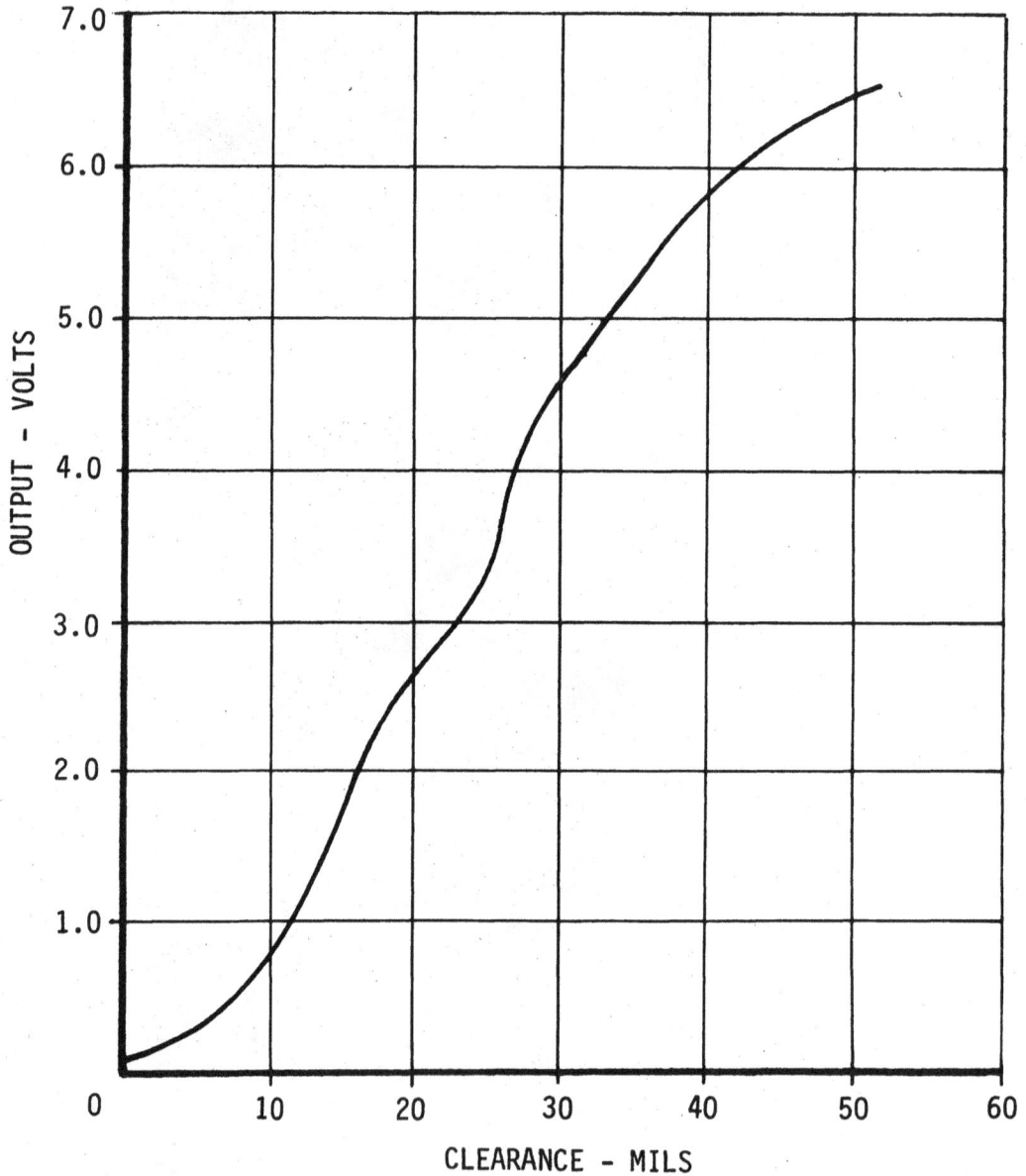

Figure 20. Preliminary System Calibration.

5.   The usable clearance signal decreased with an increase in target temperature due to a decrease in signal-to-noise ratio.

6.   It was determined that the probe could be used to measure hot clearances; however, a field stop was needed at the end of the fiber optic bundle. Figure 21 shows that the signal reflected from the 1500 F target is less than 50 percent of the total energy available. Since the total energy from the target saturates the diode above 1500 F, a means of reducing the total energy to the face of the diode was required. Placing a field stop at the end of the input fiber optic bundle would reduce the total energy falling on the diode. This would allow the diode to come out of saturation. The diode would then be able to distinguish the reflected spot from the hot target.

7.   The diode was found to be sensitive to intensity as well as to position.

8.   The depth of field was determined to be 0-65 mils.

9.   The speed of response of the diode was determined to be 5 microseconds, as shown in Figure 22.

Discussion of Results

During the preliminary testing, problem areas were identified from observations as well as from test results. Studies were also made concerning the adaptability of the system to the T700-GE-700 engine. The problem areas highlighted were:

1.   Insufficient laser power for a clearance signal.

2.   The SC-10 diode position detector's sensitivity not only to position but also to intensity.

3.   Hot target above 1400 F emits energy of the same wavelength as the laser beam.

4.   Optical sensitivity of probe assembly to stack-up of tolerances.

Changes Implemented

As a result of the preliminary testing, and consultations and discussions with outside vendors, changes were made and a complete system was built for laboratory testing. The major changes in the system were:

1.   A 15-milliwatt laser was installed in place of the 10-milliwatt laser.

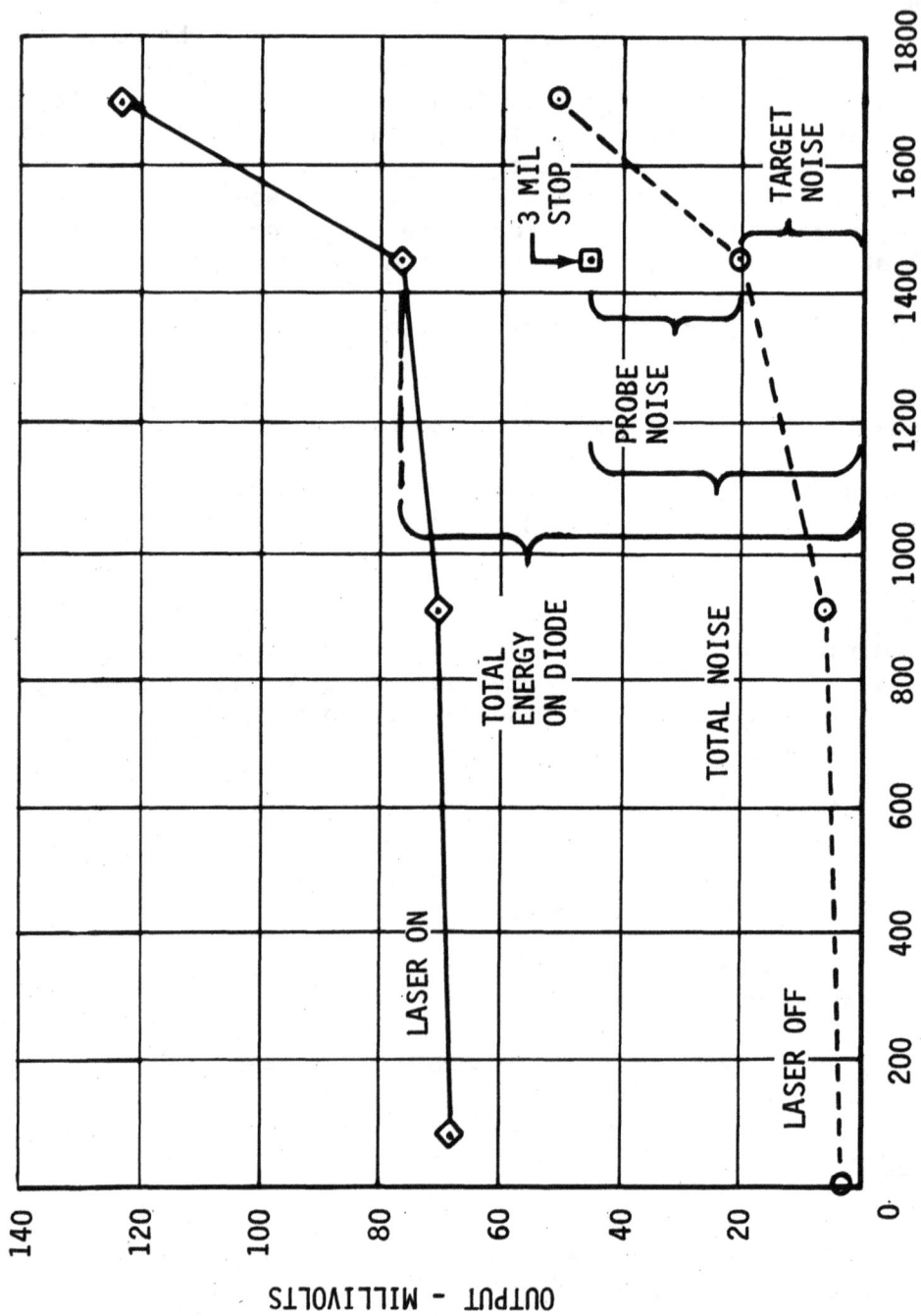

DIFFUSED TARGET TEMPERATURE - °F

SIGNAL = TOTAL ENERGY - (PROBE NOISE + TARGET NOISE) = 30 MV AT 1450°F

Figure 21. Determination of Available Signal With a Hot Target.

58

DIODE SPEED OF RESPONSE

90%

10%

5 MICROSECONDS

STEP INPUT FROM LIGHT-EMITTING DIODE

5.976 MICROSECONDS

SQUEALER
TIP

1.4 MICROSECONDS

34.986 MICROSECONDS

TIME BETWEEN BUCKETS

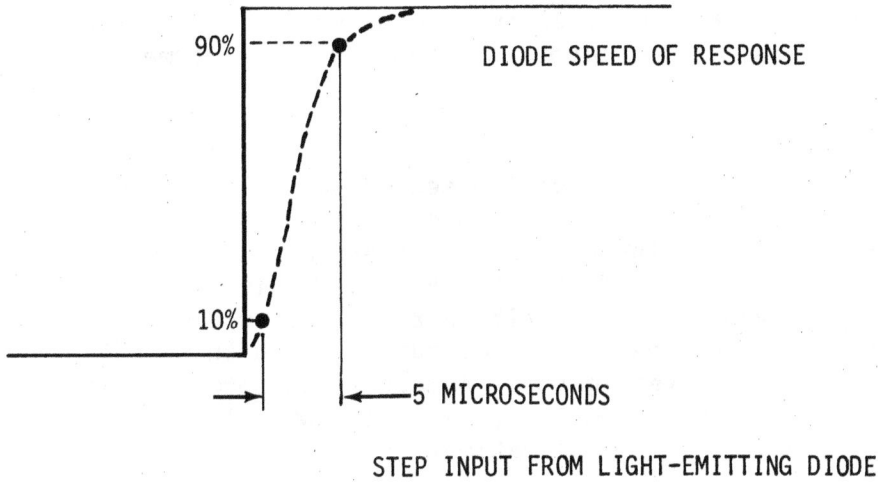

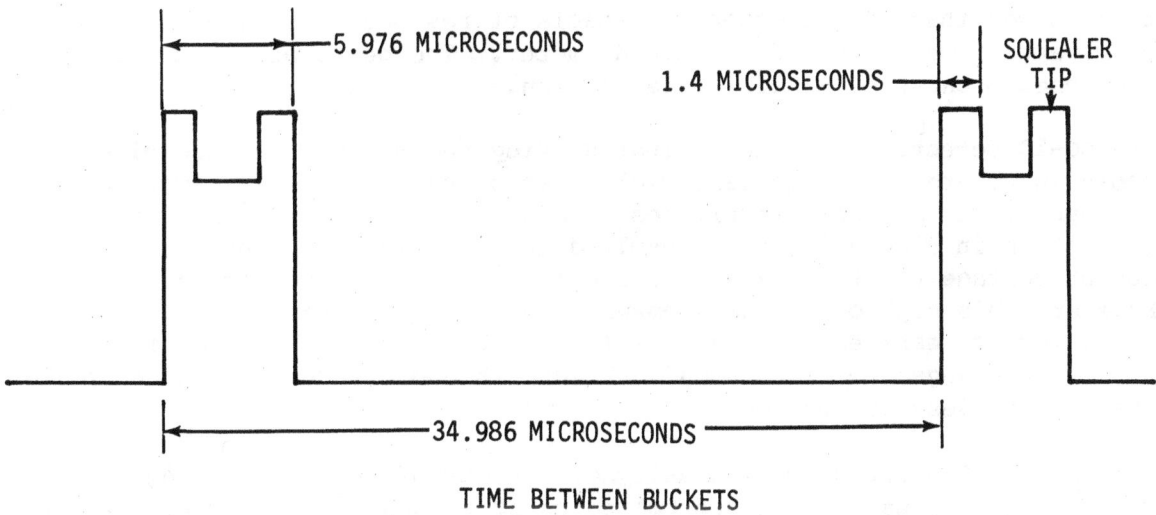

T700 BUCKET DURATION UNDER MTCMD AT 45,000 RPM

Figure 22.   Diode Response Rate and Blade Tip Requirements.

2.   A slotted field stop with a slot width of 0.030 inch was added
      at the downstream end of the output fiber optic bundle to
      reduce target generated noise.

SYSTEM DEVELOPMENT TESTS

Following fabrication of the MTCMD probes and procurement of the detector
and associated electronics, further laboratory tests were performed.

Detector Test Results

Frequency Response: The SC-10 diode position sensor was tested to determine its frequency response (rise time). This evaluation was done to
confirm the data obtained during the preliminary tests and also to determine decay time, which was not considered at that time. Figure 23 shows
both the rise and decay times. Fast repetition rates had no significant
effect on the performance of the detector. Concurrent with this test
another position detector was investigated. Although the specification
indicated a faster response time, laboratory testing showed it to be no
better than the model SC-10 in this application.

Linearity: The linearity of the SC-10 detector was evaluated to
determine its response a constant intensity light source. The laser
output beam was passed through a 10-mil aperture to simulate the
reflected output spot size which would impinge on the detector. Beam
position on the diode versus output volts is shown in Figure 24. This
test showed that the detector is capable of responding to a small-size
beam, approximately 0.010 inch in diameter. The peaks of the curve are
flattened because of amplifier saturation.

The SC-10 detector was then evaluated using the MTCMD using the circuit
shown in Figure 25. This test yielded an output voltage that was low and
nonrepetitive, as shown in Figures 26 and 27. The circuit was changed to
that shown in Figure 28 which resulted in a considerable increase in
output voltage for full target deflection (see Figures 29 and 30).
However, this high output was somewhat nonlinear but satisfactory and
confirmed the earlier test results (Figure 20). Note that the circuit
was later changed to that shown in Figure 18 and discussed further in the
System Test Results section.

Intensity: The effect on intensity of the output beam with change of
distance of the target was determined using an unpolished nichrome target.
The results of the test using the unpolished nichrome target are shown in
Figure 31. The target displacement which causes the reflected beam to
diffuse does not affect the output until it is beyond 0.025 inch. The
use of both the cutaway and the full probes confirms this characteristic.

The intensity changes clearly affect the output of the SC-10 detector as
shown in Figure 32. This intensity change may be caused by many factors;
the most important being the reflectivity of the target material.

60

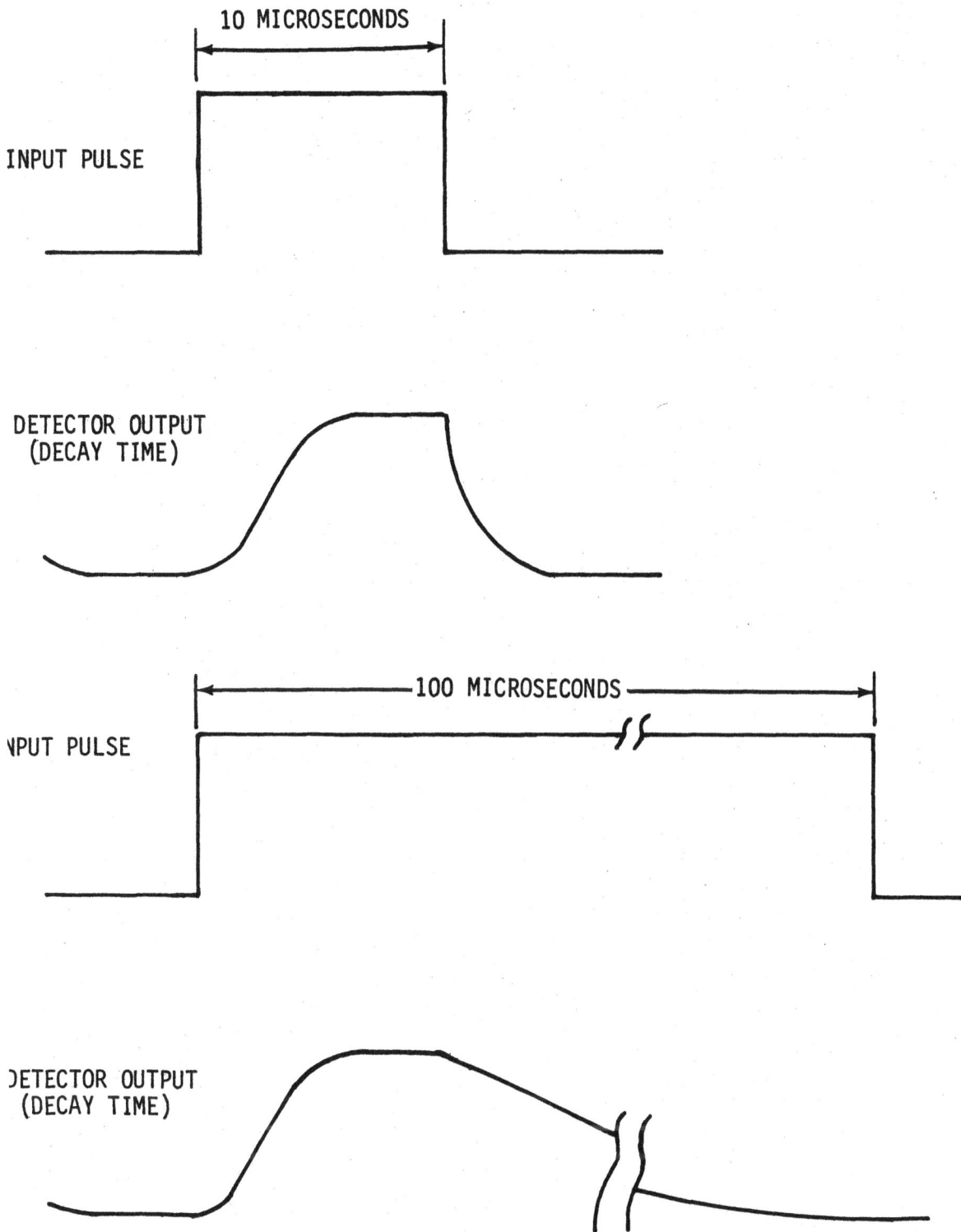

10 MICROSECONDS

INPUT PULSE

DETECTOR OUTPUT
(DECAY TIME)

100 MICROSECONDS

INPUT PULSE

DETECTOR OUTPUT
(DECAY TIME)

Figure 23.   Decay Rate of SC-10 Detector.

61

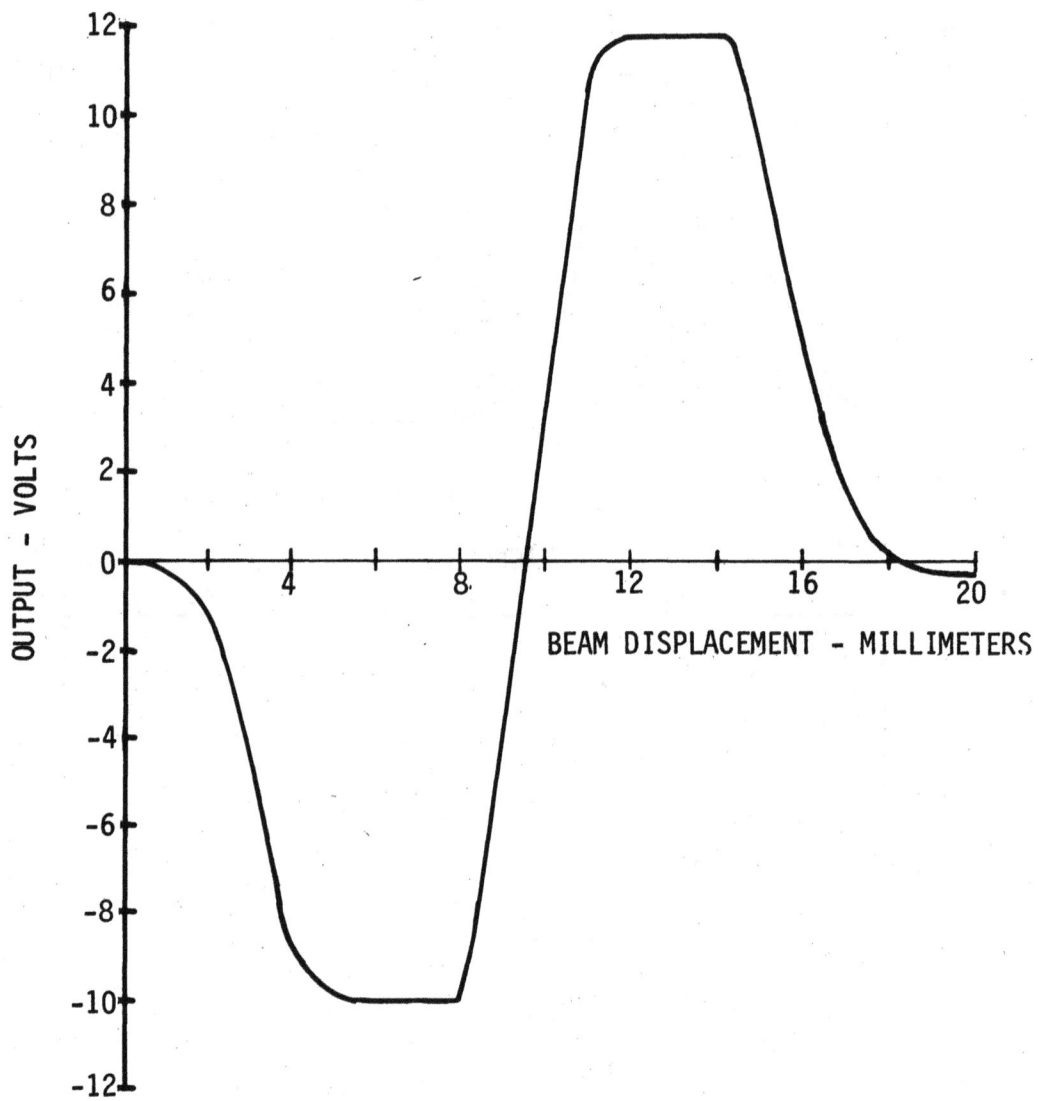

Figure 24. SC-10 Linearity Across Sensitive Area.

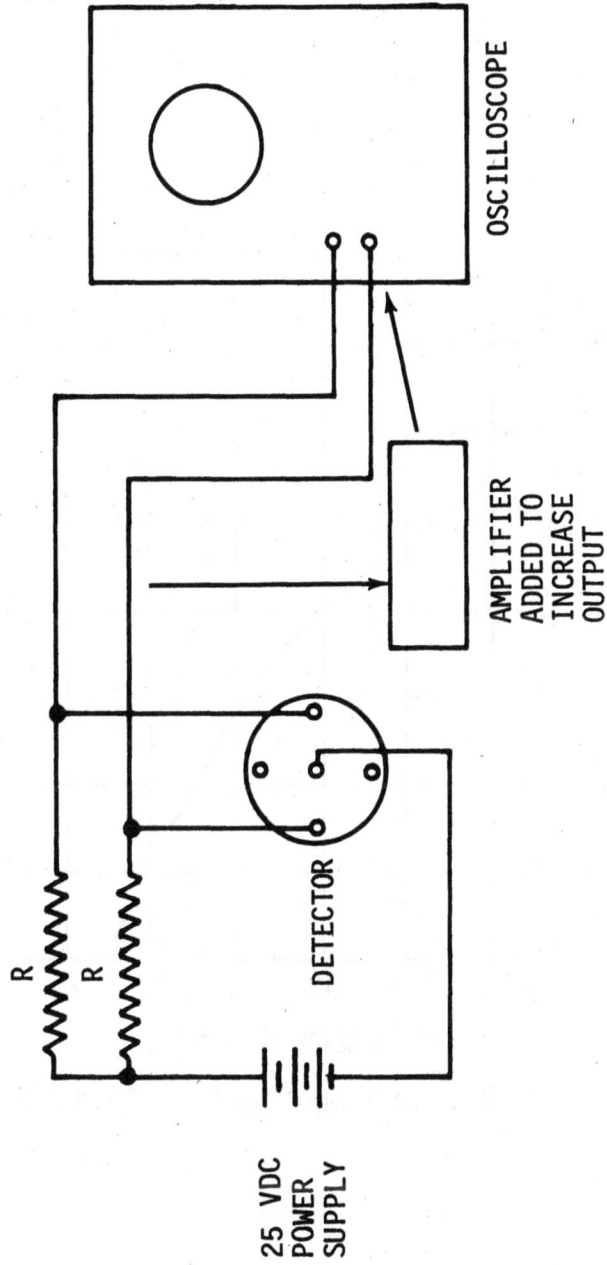

Figure 25. Detector Circuit Recommended by Manufacturer.

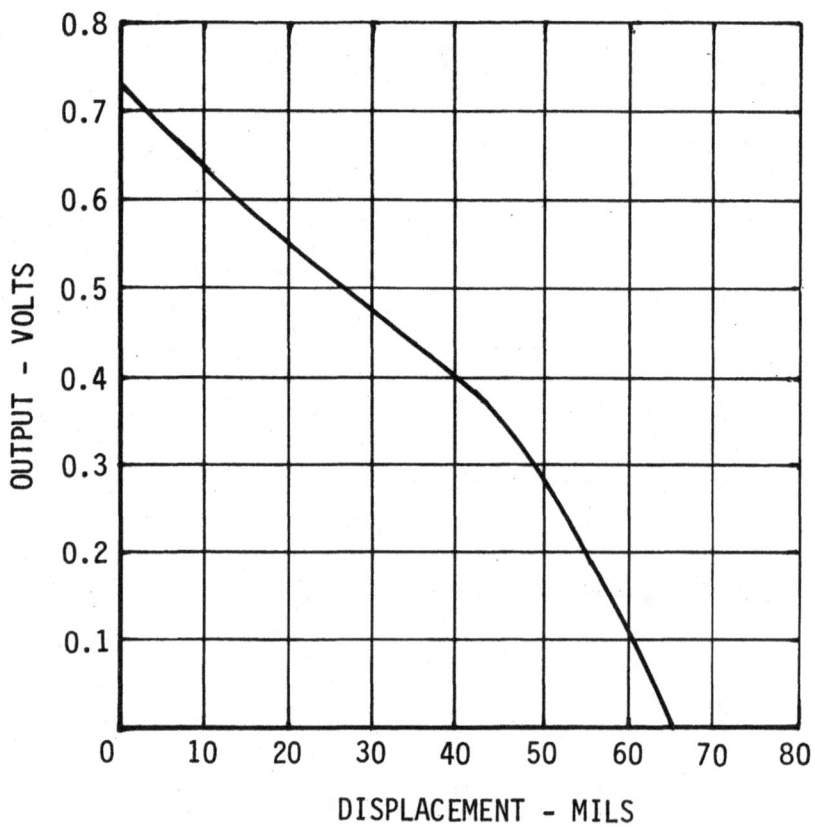

Figure 26. Calibration of Detector Circuit.

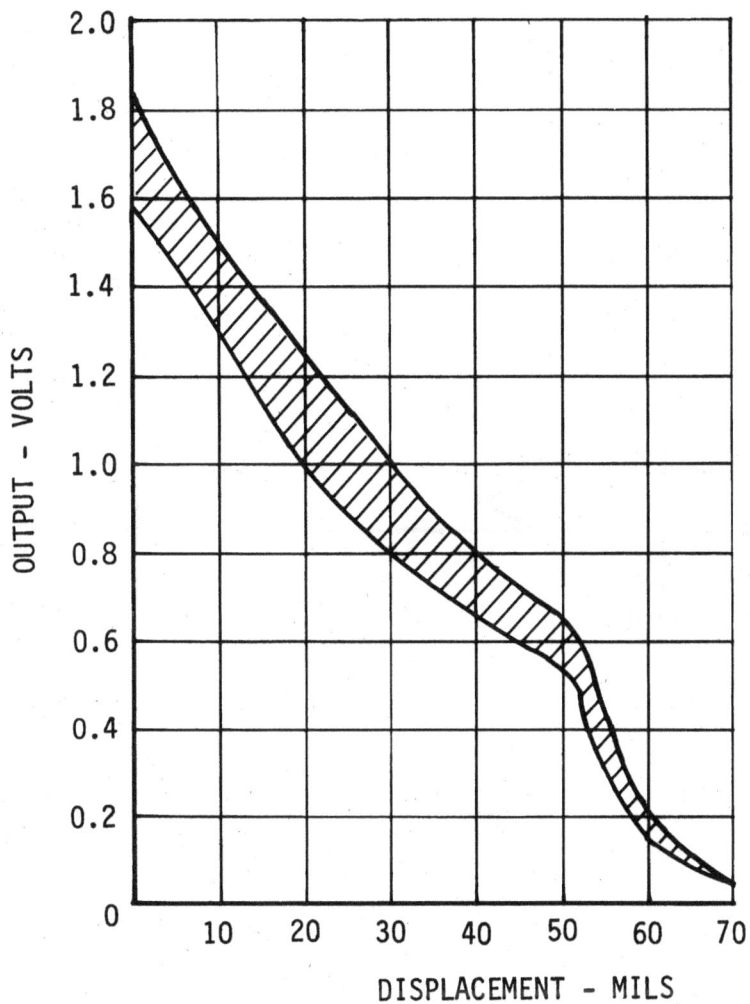

USING DETECTOR CONNECTIONS IN FIGURE 25 USING AMPLIFIER

Figure 27. Calibration of Detector Circuit.

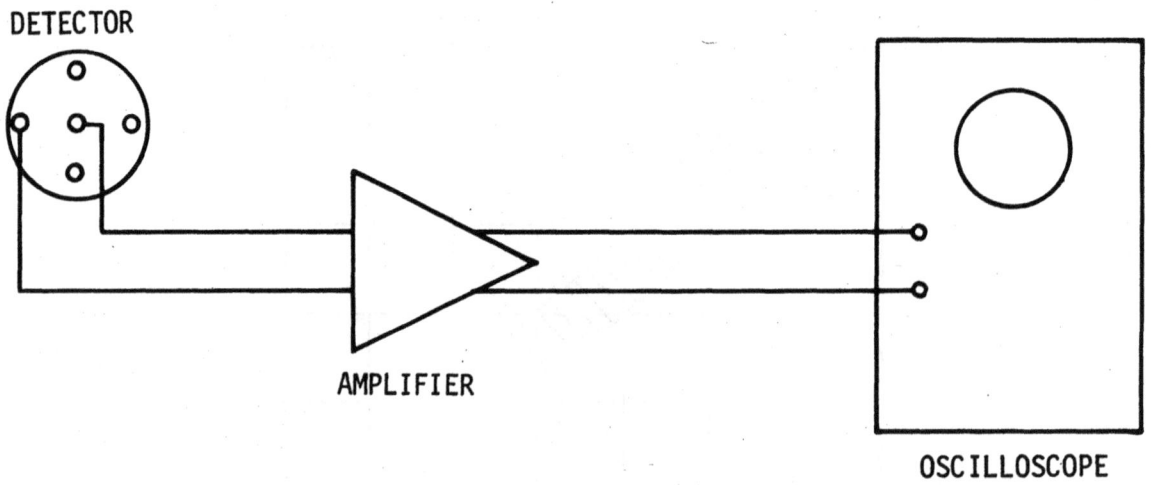

DETECTOR

AMPLIFIER

OSCILLOSCOPE

Figure 28. Detector Used to Obtain Highest Output.

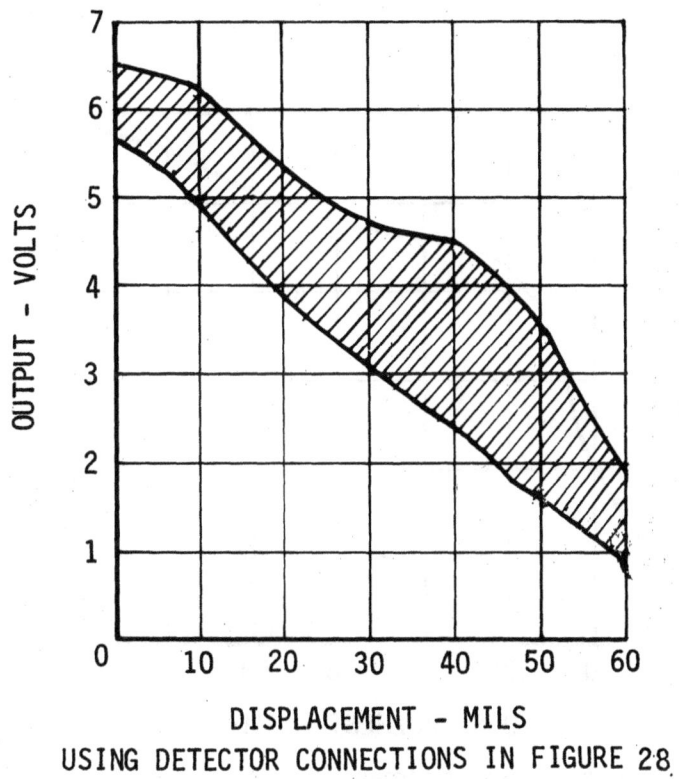

Figure 29. Calibration of Detector Circuit.

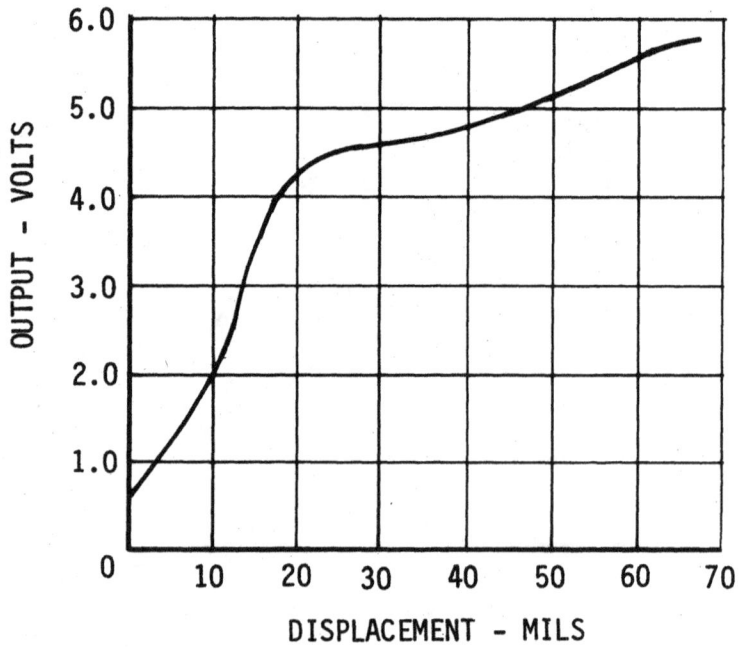

USING DETECTOR CONNECTIONS IN FIGURE 28 REVERSING OUTPUT

Figure 30.   Calibration of Detector Circuit.

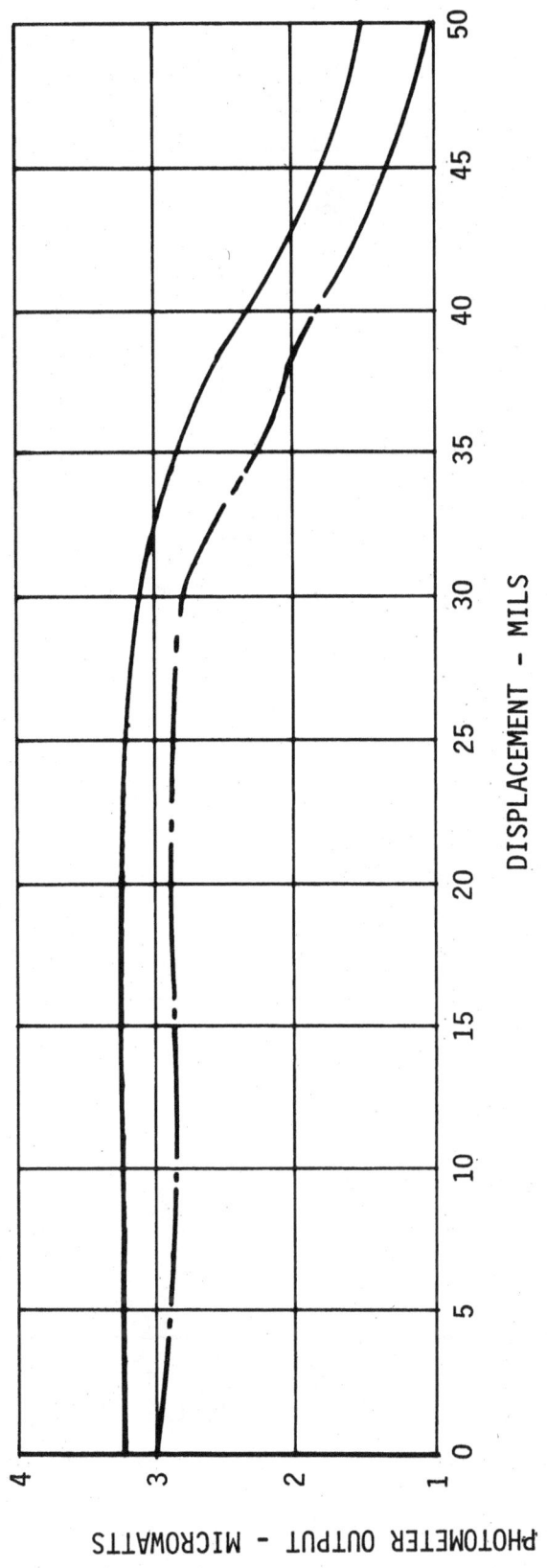

Figure 31. Effects on Intensity vs. Target Displacement.

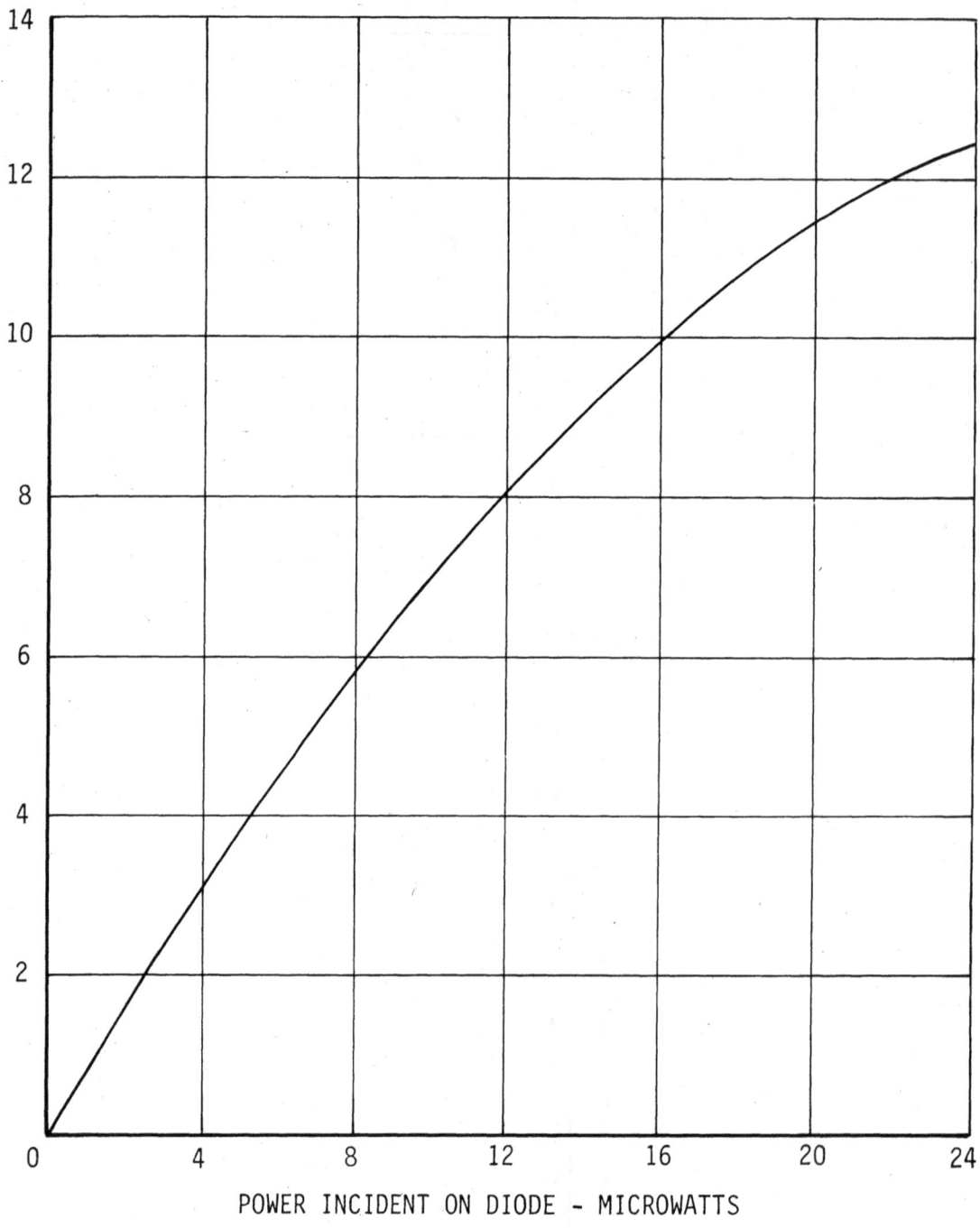

Figure 32. Effect of Beam Intensity on SC-10 Detector.

In addition, Figure 33 shows the relationship between output voltage of the detector and the incident intensity in microwatts. This data was taken with the circuit shown, where the laser beam was incident on the face of the detector and attenuated in various steps by using optical filters.

A third method of measuring the effects of intensity is shown in Figure 34. Here the test was the same as for Figure 33 except the current-to-voltage amplifiers were removed and the signal was offset from the center of the diode.

These tests indicate that the detector is very sensitive to intensity at low-power levels. As the power increases, the sensitivity is diminished as shown by the flattening of the curves in Figures 33 and 34.

Hot Target: The MTCMD was evaluated for target distance detection using a heated target. The results of these tests are shown in Figure 35. Each plot is at a different target temperature, beginning at 400°F and progressing to 1400°F. As recorded in the preliminary tests, the detector becomes insensitive to the reflected laser beam at temperatures above 1400°F due to the saturation caused by infrared radiation and indicating that a band-pass filter is required.

Optical Tests Results

Fiber Optics: The uniformity of fibers in the output fiber optic bundle was determined with the results plotted in Figure 36. This plot shows the transmission to be nonlinear but symmetrical about the center of the bundle over the mid-portion of the bundle face.

Optical power loss in the system was measured and is shown in Table 8. About 20 percent of the power is lost through the microscope objective; 58 percent of the remaining power is lost through the input bundle. Using an unpolished target, the optical power output is about 0.1 percent of the laser power. With a highly polished target, the output power is about 7.5 percent of the laser power and the aluminum foil target yields about 1.3 percent of the laser power.

Vibration Sensitivity: During the evaluation of the SC-10 detector using the MTCMD, a problem became apparent when the setup was disturbed and the high signal output was lost and could not be regained. While searching for the cause it became apparent that the intensity of the beam was fluctuating considerably. To provide better observation, the split probe was used and the beam was observed while moving the fiber optic bundles. It was determined that the input fiber optic bundle had broken fibers, which created the change in intensity.

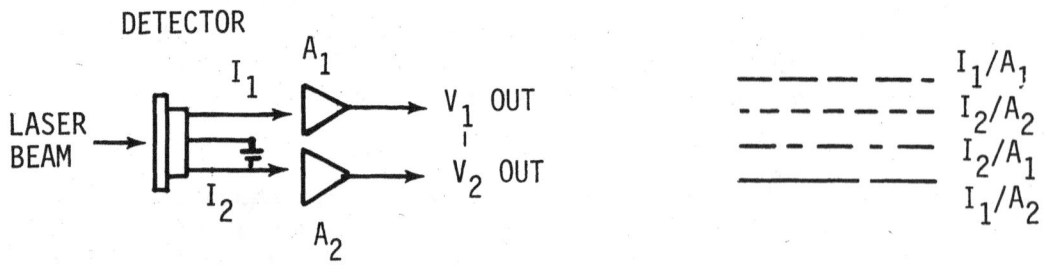

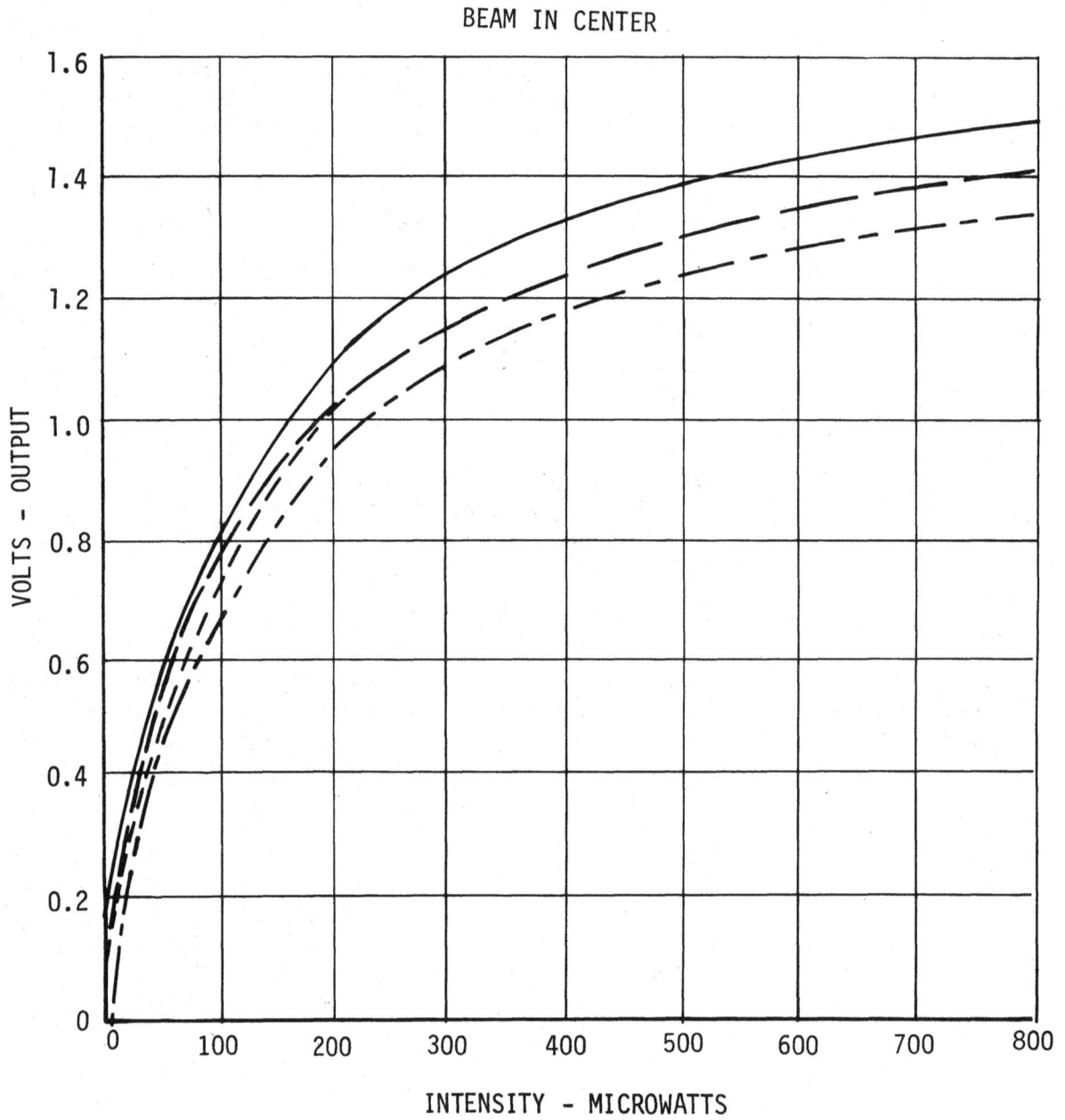

Figure 33. Detector Voltage Output.

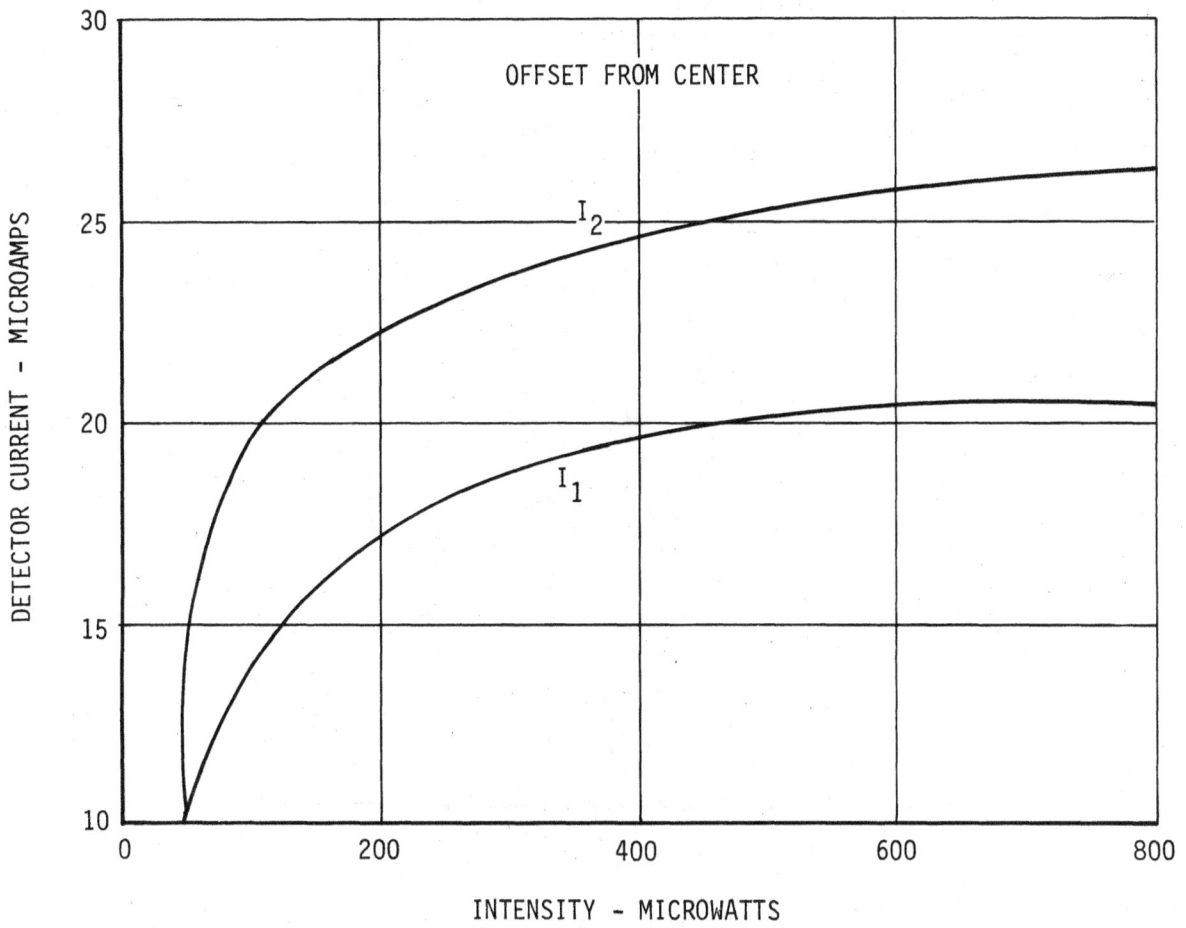

Figure 34. Diode Current Output Calibration.

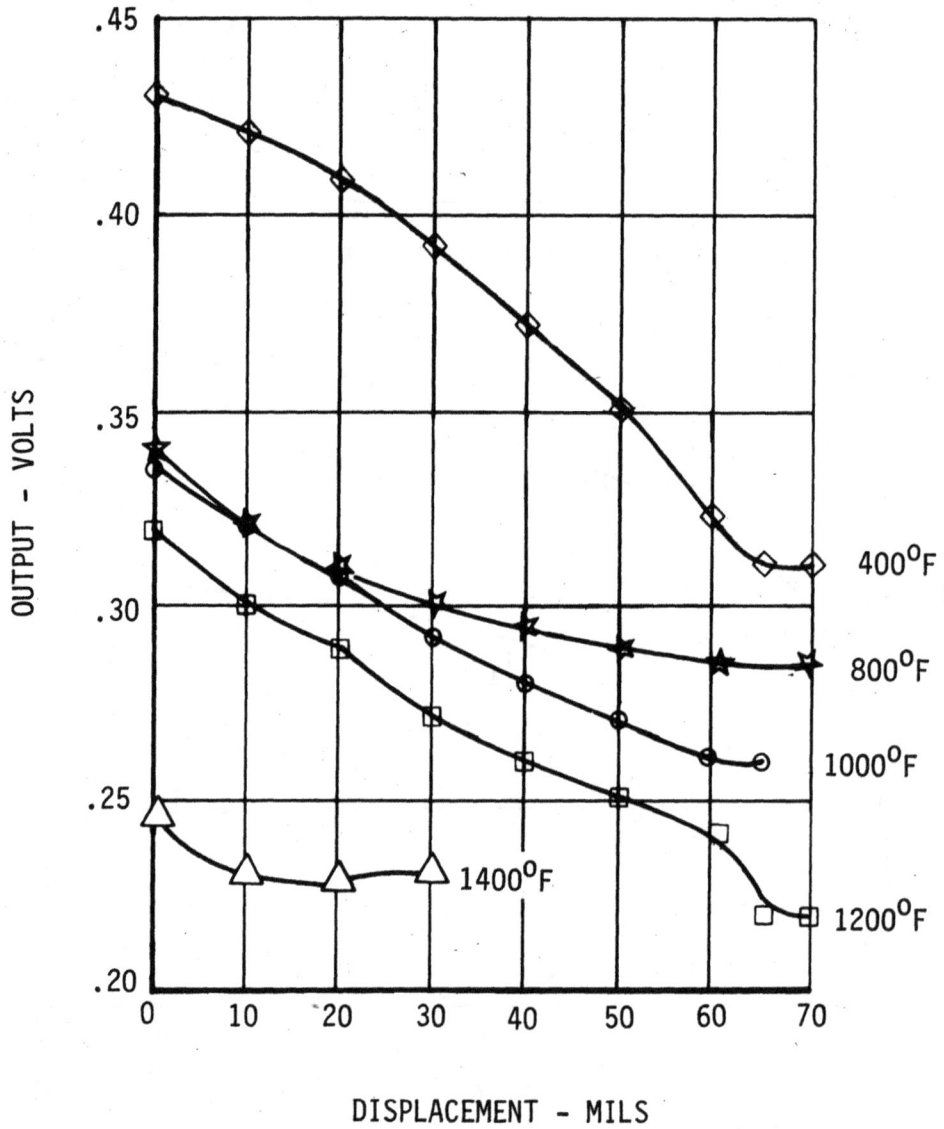

Figure 35. Output Signal Calibration With Heated Targets.

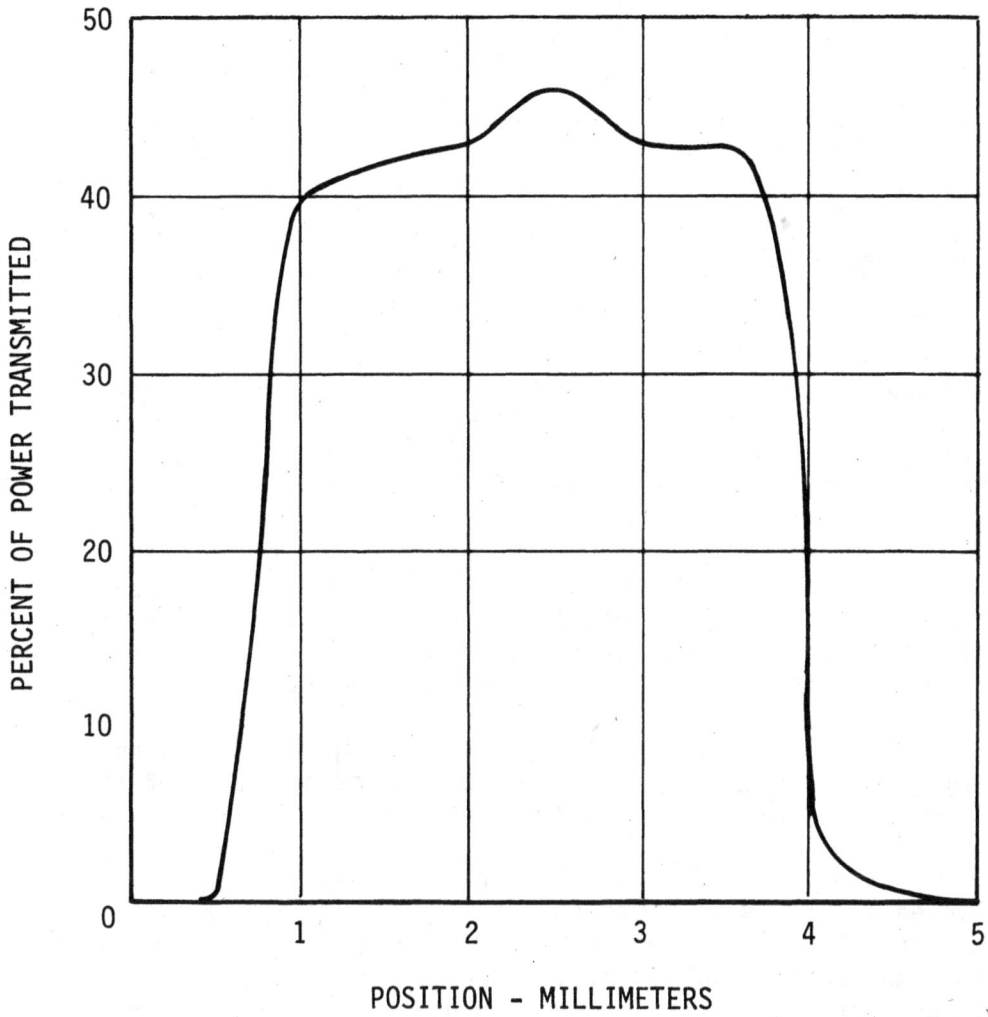

Figure 36. Output Bundle Transmission Efficiency.

Equipment	Output (Microwatts)
Laser	18,000
Microscope Objective	15,500
Input Bundle	5,500
Probe Output	1,500
Output Bundle	
a. Unpolished Nichrome	80
• Filtered	15
b. Shiny Metal (SST)	100
• Filtered	20
c. Unpolished SST	35-40
• Filtered	10

TABLE 8. OPTICAL POWER LOSSES

After replacing the input fiber bundle, the remainder of the system was checked thoroughly. The full probe was reinstalled and the system was again checked. The output voltage was quite low, which was attributable to the critical alignment required for the fiber bundle.

It was also observed that close attention must be given to ambient vibration, as small increments of motion (thousandths of an inch) of the fiber bundles affect the output considerably.

Noise: While testing with the cutaway (split probe), a high level of stray reflected light was observed and the reflective surfaces of the probe were coated in a dull black finish. By sharing a lens and prism between input and output functions, the design causes a great deal of light to be scattered directly from the input (source of illumination) to the output (face of fiber optic output bundle) without being reflected from the blade tip as intended. Major sources of scattered light are the front surface of the probe lens and the prism. This stray light gives a steady bright background against which the clearance signal must compete for detection. The bright background makes position detection difficult for any type of sensor. In practice, the background light fluctuates as an unknown function due to vibration of the system, while the actual clearance signal is pulsing rapidly at the blade passing frequency.

A real-time detector such as the SC-10 diode will see the rapid pulsation of the clearance signals (each at its proper location) against a more slowly varying, but symmetrical, background of light (noise). Since the position sensor responds to the centroid of the instantaneous light distribution, the noise weights the response toward the center of the diode, effectively reducing sensitivity to position information.

Reflectivity: Two cases of reflection from blade tip surfaces are important. The blade tip may be highly polished so that the rays of light leaving the probe tip are nearly all reflected at the specular angle (angle of incidence equals angle of reflection). The MTCMD is designed to capture rays reflecting from the blade tips at this angle. However, the specular reflectance of materials is very strongly affected by their surface condition. A slight amount of roughening or deposit will cause the reflected rays to leave the blade tip at other angles, and not contribute to the collected clearance signal. An unpolished surface represents the other extreme case of reflection, i.e., diffuse reflection. In this case, the incident beam is reflected at nearly all angles, and very little light falls within the angle of acceptance of the probe. Setting the probe optical geometry for reflection at the specular angle of reflection results in higher initial outputs, but is far more sensitive to surface condition than a diffuse angle since most materials have a greater spread in specular reflectance than in diffuse reflectance from one material to another. Diffuse reflection, while less sensitive to surface condition, is very inefficient as little incident light is scattered in any one direction.

In subsequent laboratory and engine tests, various blade tip coatings such as platinum and CODEP were tried without success in order to provide a more reflective finish with greater resistance to darkening (oxidation) during engine operation.

Infrared Filtration: Laboratory tests with heated test targets also led to the requirement of a laser band-pass optical filter in place of the interference filter specified for the original MTCMD design. Blade tip radiation in the visible and infrared wavelengths creates an even illumination of the output bundle against which the desired clearance signal must compete for detection. Unlike the stray light, which is scattered within the probe and which is relatively constant, the target radiation has the same periodicity and duration as the clearance signal. The only other contribution at this frequency is a possible addition to the background scattered light from rays reflected from the blade tip at an angle which re-enters the probe and is scattered onto the output bundle.

At low temperatures, below 1100°F, blade tip radiation is predominantly in the infrared region and only at higher temperatures is the radiation visible, with a small portion having the same wavelength as the laser light. An optical band-pass filter is very effective in screening out light not at the laser's wavelength. Typically, less than 1.5 percent of the bright red radiation from a hot source falls inside the band-pass of a high quality narrow filter. However, the specifications for the filter on the original MTCMD were to block the lower (below 632.8 nanometers) wavelengths of the infrared spectrum. The original filter was replaced by one of a more effective design, but at the cost of more attenuation of the laser beam.

A narrow band-pass filter transmits 40 percent of the laser energy incident upon it and less than 1.5 percent of infrared radiation energy below 2000°F. Since these filters are interference filters, they are very sensitive to the direction of the incident light. If a ray of light at the design wavelength passes through the filter at an angle to the normal, it will "see" a filter with a longer effective central wavelength and will be attenuated. Thus the effective transmittance of laser light is thereby reduced from 40 to 25 percent or less. The original MTCMD design evidently did not consider reducing the divergence of the light, leaving the end of the output fiber optic bundle before it strikes the optical filter. This characteristic also applies to the narrow band-pass filter utilized for the GE System. The test results are shown in Figure 37.

System Test Results

In order to determine the threshold of detector operation versus available power, the complete system with the electronic conditioning package was evaluated.

The earlier data indicated that the detection system required approximately 200 microwatts (MTCMD output) to function effectively. It was anticipated, based on measured losses through the system, that only 30 to 100 microwatts would be produced under engine operating conditions. Systems operation at that low level is susceptible to electrical, acoustic and vibration noise interference. Consequently, an electronic circuit was devised to minimize voltage changes due to intensity and to raise the output of the detector to a usable voltage. The circuit is a current to voltage and voltage divider circuit as shown in Figure 18.

The SC-10 position sensor is provided with two electrical output circuits. The current resulting from the light input is split proportionally to the position of the impinging beam relative to the center of the sensor. Measurements showed that the two currents were proportional to intensity as well. However, the ratio of the two currents is relatively independent of the intensity.

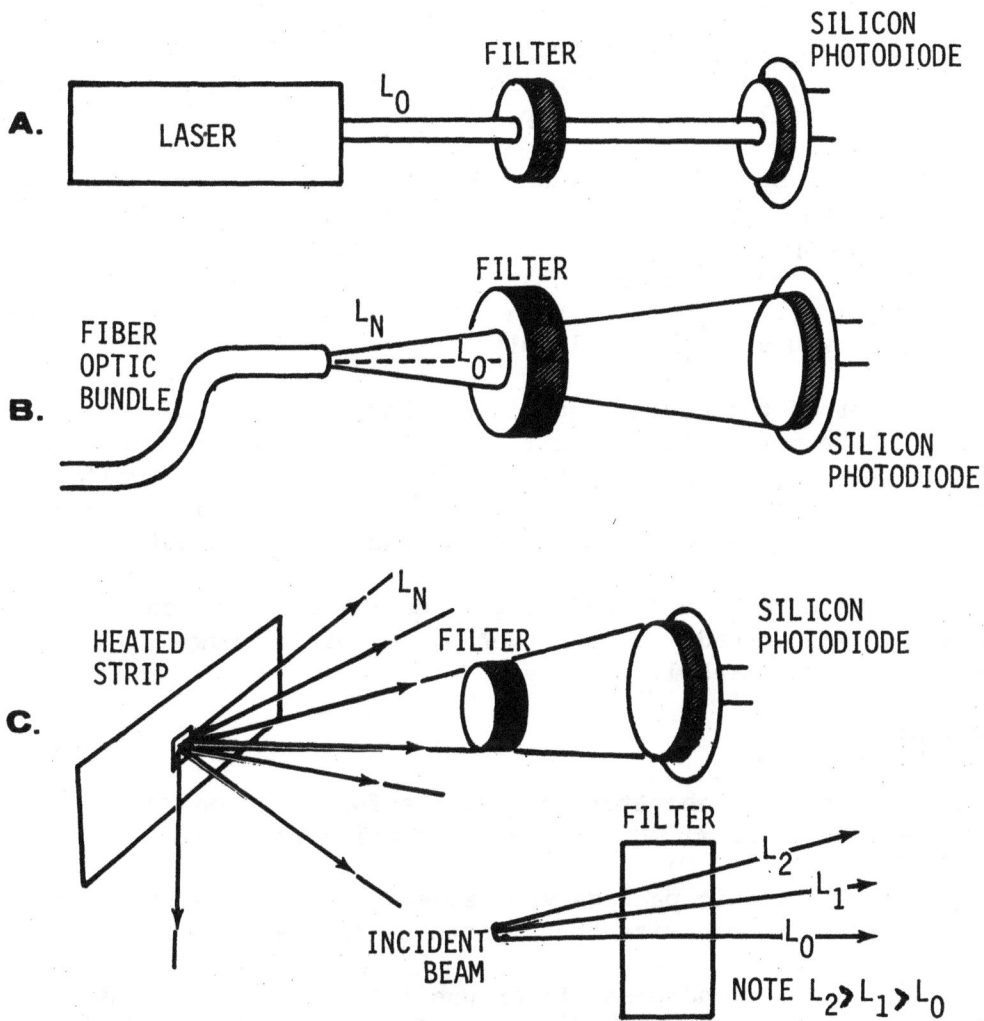

A. FOR COLLIMATED LIGHT, 632.8 NANOMETERS (40%)

B. FOR DIVERGENT LIGHT, 632.8 NANOMETERS (25%)

C. FOR DIVERGENT LIGHT, 400 to 2000 NANOMETERS
   (25% FOR STANDARD INTERFERENCE FILTER)
   (1.5% FOR NARROW BAND-PASS FILTER)

Figure 37. Efficiency of Band-Pass Filter.

The theory of operation begins with the two currents from the diode which are converted to usable voltages in the first two amplifiers Al and A2. Then amplifiers A3 and A4 form the difference and sum of these respective voltages. An analog divider module takes the ratio of these two signals. A line driver feeds the clearance indication (now a voltage) to CRT and recording equipment.

The first-stage amplifiers must supply very high gain. With 0.1 megohm feedback resistors, they convert current to voltage at the rate of 10^5 volts/amp or 100 millivolts/microamp.

Theoretically, the sum of the two derived voltages is proportional to the total optical power incident on the diode. Figure 38 shows that the relationship between optical power and diode current (hence, voltage after amplification) is nonlinear, and that even with an ideal polished target, the system would deliver no more than 500 millivolts to the divider. Typical signals would be 100 millivolts or less.

The manufacturer of the divider specifies that the magnitude of the denominator voltage be more than one volt. This type of divider has been observed to operate well with as small a signal as 200 millivolts. However, it is undesirable to work in a region where small fluctuations in the input cause large changes in the already small output signal.

Many attempts were made to raise the gain of the electronics; however, since the gain was already so high over a very wide bandwidth, each gain increase compromised stability.

Changes Implemented

Upon completion of the laboratory tests, the following changes were made to the three additional MTCMD systems to be built for engine testing:

1.   The electronic package was expanded to include the divider network to minimize the effects of intensity variations.

2.   An optical band-pass filter was incorporated to separate the hot target radiation frequency from the clearance signal (laser beam frequency).

Bench testing and analysis had indicated that the blade tips, due to the tip cutout, would present two surfaces to the laser (each 0.054 inch wide) and would allow the probe to see the bucket at each edge for 1.4 microseconds at 42,000 rpm (refer to Figure 22). Since the diode response time is 5 microseconds, the full response of the diode would not be realized and dynamic calibrations would be required to correlate the data. Five blades were fabricated with solid tips which would give a passing duration of 5.98 microseconds and not require dynamic calibration.

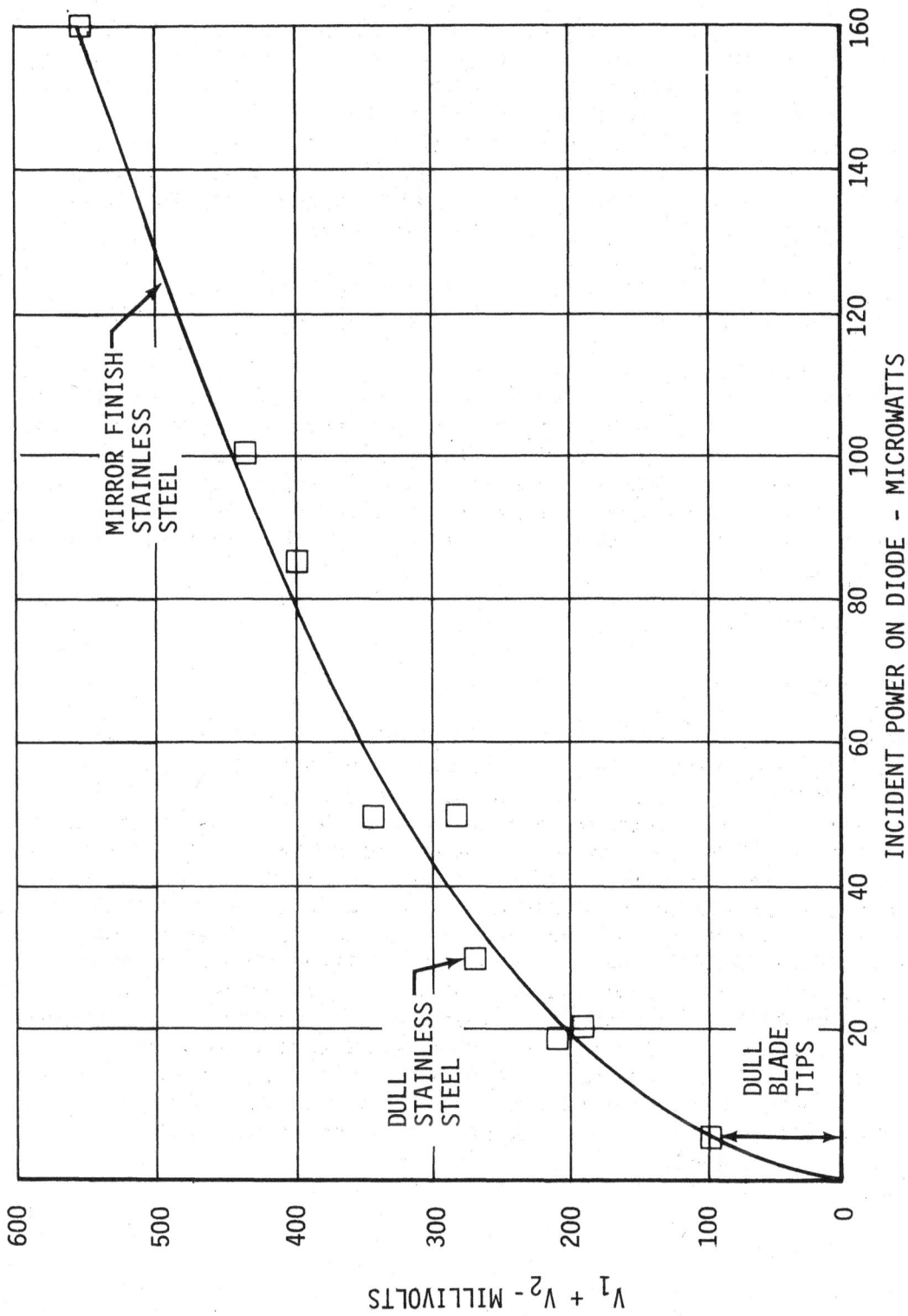

Figure 38. Amplifier Response to Optical Power Incident on Diode.

81

CALIBRATION TESTS

Prior to installation on the engine, several systems were assembled for calibration and further testing. A typical system is shown in Figure 39. The calibration was carried out on an optical bench. Both probe and target were rigidly mounted on carriers, and clearance was set with a lead-screw and precision dial indicator. This configuration allowed precise measurements of system accuracy and repeatability. Additional benefits were that the target material could be easily changed and the effects of misalignment and electronics drift easily observed.

Tests in the laboratory showed the systems output signal to drift in the absence of a reflective target. Cooling of the electronics reduced the drift.

Typical calibration curves of volts output versus clearance are shown in Figures 40 and 41. Repeatability of the calibration curves led to the belief, which was later verified experimentally, that the system was working with the very lowest intensity levels at which it could function. Calibration tests demonstrated that the calibration curve of volts versus mils of clearance is linear and not affected by blade surface reflectivity (intensity changes).

ENGINE TEST RESULTS

Engine Test No. 1

Based on the bench test data which indicated a marginal reflected signal strength, it was decided that for this checkout one of the two installed systems (6:30 o'clock) would not contain the optical bandpass filter. A typical engine installation is shown in Figure 42. When the engine was run at Idle (28,850 rpm), the oscilloscope connected to the 6:30 system showed high-frequency oscillations. However, the 12:30 (filtered) system showed pulses indicative of clearances. The signals were recorded on magnetic tape and Polaroid pictures taken of the oscilloscope display. Figure 43 is representative of those taken. Although pictures and recordings were made, it was noted that the direct current reference was unstable and drifted extensively. This drift prevented a valid interpretation of the signals with respect to actual clearance measurements. In addition, the polarity of the signals was reversed from that of the calibration. The pulses indicated clearly when a solid tip blade was passing the probe while the split blade signals were not as well defined as the solid tip blade signals.

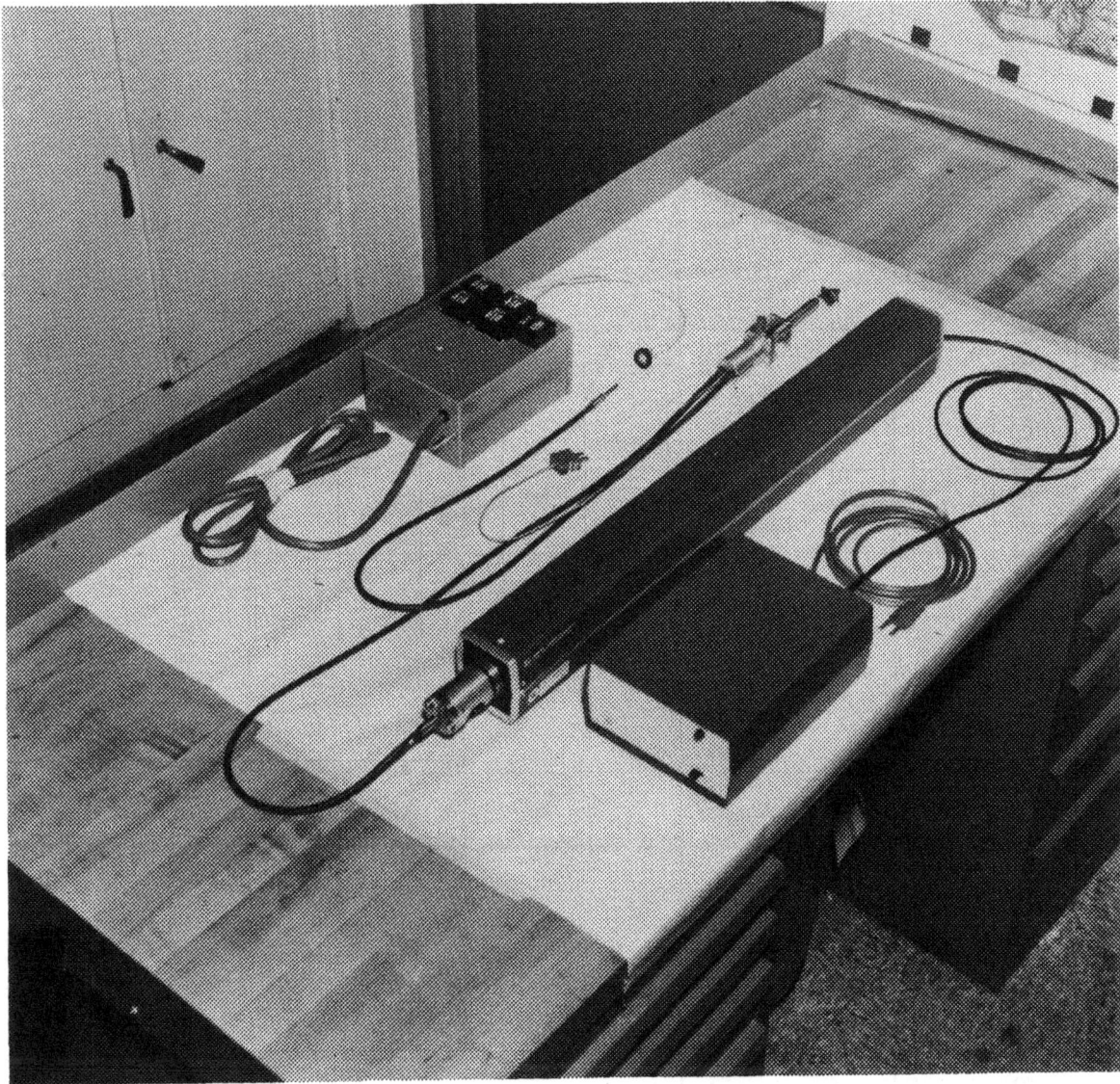

Figure 39. Turbine Tip Clearance Measuring System.

83

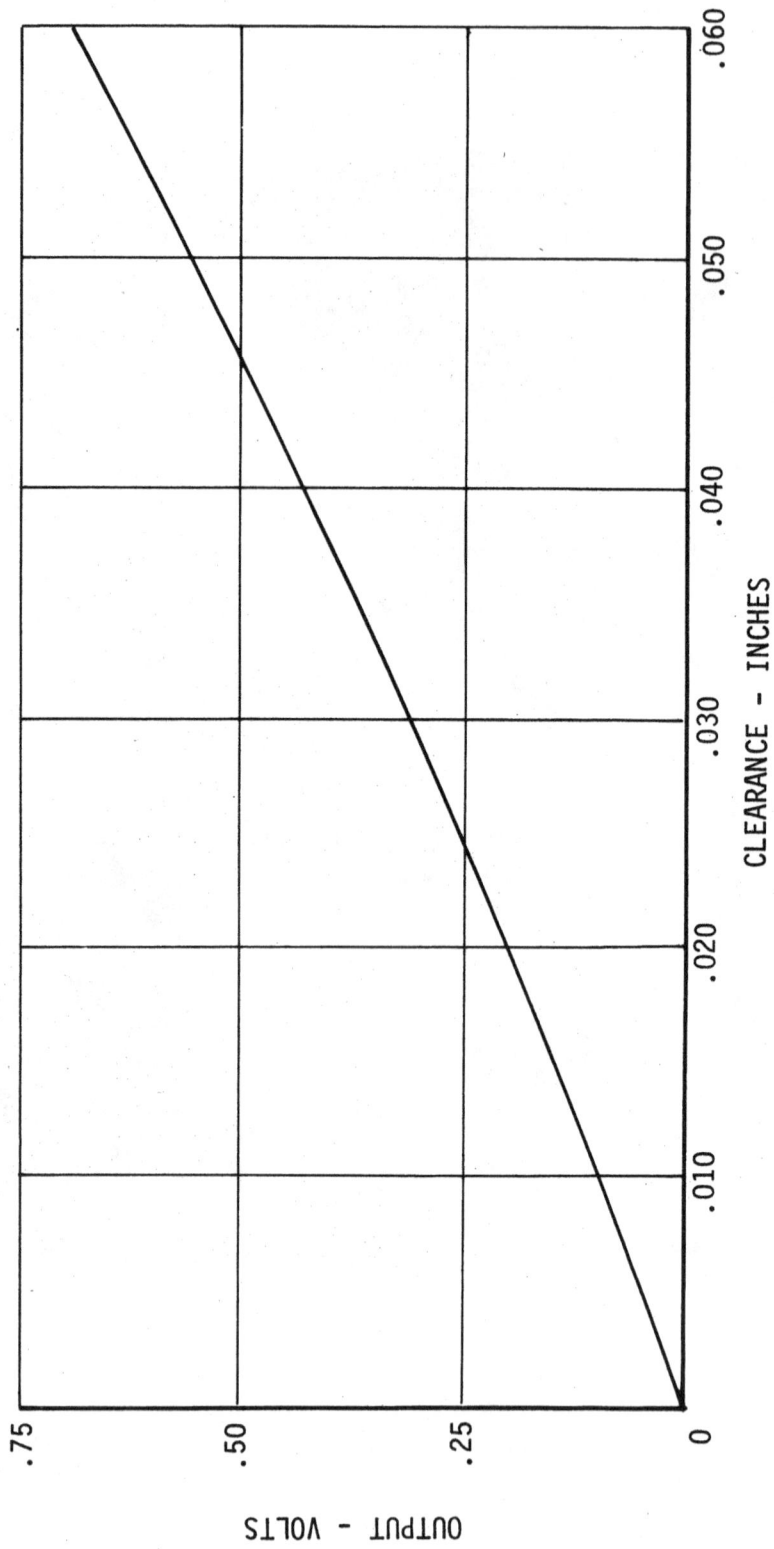

CALIBRATION CURVE TURBINE TIP CLEARANCE MEASUREMENT SYSTEM AT 6:30 O'CLOCK

CLEARANCE – INCHES

OUTPUT – VOLTS

Figure 40. Calibration Curve Turbine Tip Clearance Measurement System at 6:30 O'Clock.

84

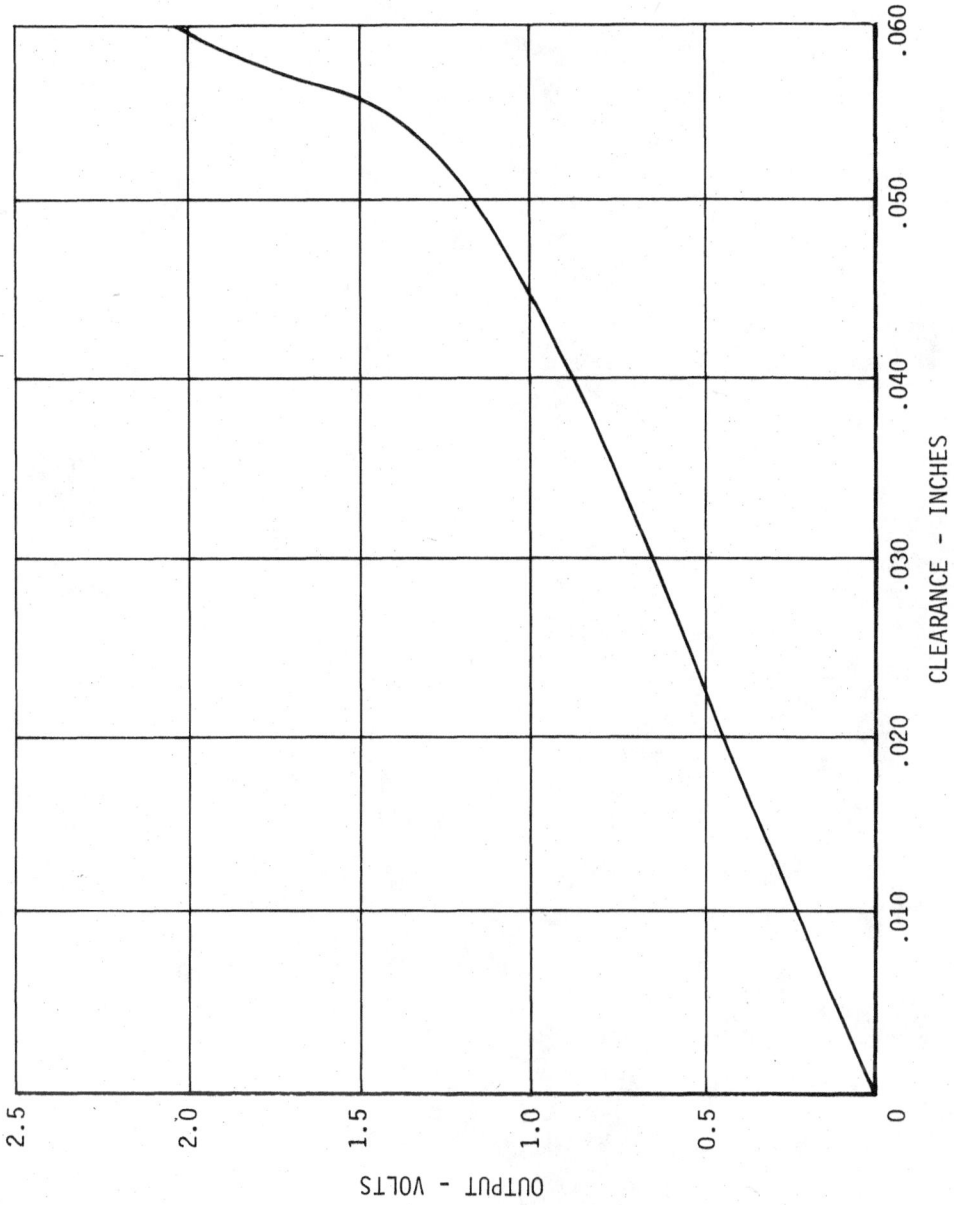

CALIBRATION CURVE TURBINE TIP CLEARANCE
MEASUREMENT SYSTEM AT 12:30 O'CLOCK

CLEARANCE - INCHES

OUTPUT - VOLTS

Figure 41. Calibration Curve Turbine Tip Clearance Measurement System at 12:30 O'Clock.

Figure 42. Typical Engine Installation Without Optical Band-Pass Filter.

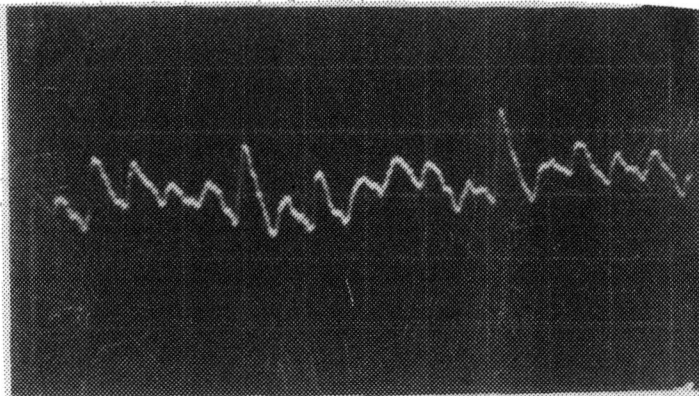

Figure 43.   Output of TTMC Without Optical Filter - Engine at 28,850 RPM.

Following the initial test, further investigation proved that the long transmission cables between the engine-mounted electronics and control room readouts caused the amplifiers to oscillate and reversed the polarity of the signal when the oscillations were not present. When the signals were monitored in the test cell with a short cable, the polarity of the signal was correct and the high-frequency oscillations disappeared. The engine was rotated manually to obtain signals while performing these checks.

In summary, the problems were:

1. Reversal of signal or polarity.

2. Amplifier oscillation.

3. No usable signal from "filtered" system.

4. Inoperable touch probe.

5. Drift of direct current zero reference level.

Items 1 and 2 were corrected by installation of a line driver amplifier. Item 3 was found to be due to a crimped fiber optic bundle. The touch probe was binding as installed, and this was also corrected.

The drift of the direct current zero reference could not be corrected immediately within the constraints of the existing measurement system and electronics. An alternating current coupled system was devised that was unaffected by drift. However, this system requires a known clearance reference at any given condition. It was decided to utilize the touch probe to provide the reference for the second run.

Test No. 2

For the second test run, the clearance probes were reinstalled, one with the infrared optical band-pass filter. The engine was rotated manually and all systems were operative. Rollover on the starter also produced good signals. The engine was started, the laser signals were observed to be degenerating and at 34,000 rpm the signals were lost. Figure 44 shows the output signal while the engine was being rotated manually and while the engine was operating. During manual rotation, three exposures were made on the film a few seconds apart with the zero reference shifted on the scope to show passage of different blades (Figure 44).

An investigation was made to determine the cause of signal loss. Upon rotating the engine manually, it was noted that the initially good signals were not present. Removal of a probe indicated that the prism was clean; good signals were obtained by aiming the beam at a highly reflective surface. A borescope inspection of the blade tips indicated a degradation of the blade tip finish, a condition that would reduce the intensity of the reflected laser beam, thereby increasing the noise to signal ratio and making the system inoperative. One of the five solid blade tips was repolished and a good signal was obtained when the rotor was rotated manually.

88

ENGINE MANUALLY ROTATED

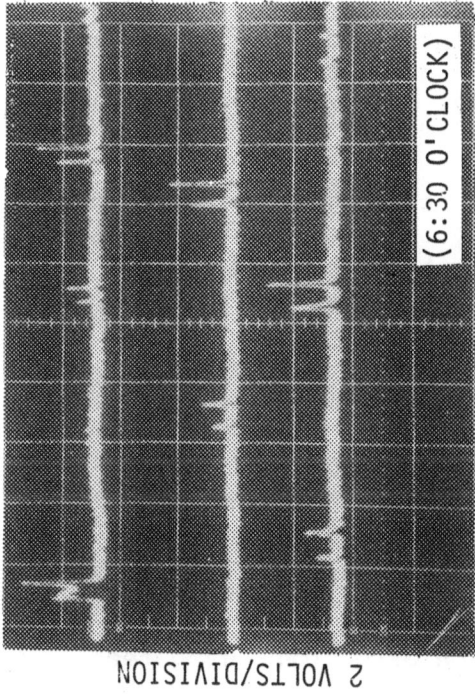

2 VOLTS/DIVISION

(6:30 O'CLOCK)

5 MILLISECONDS/DIVISION

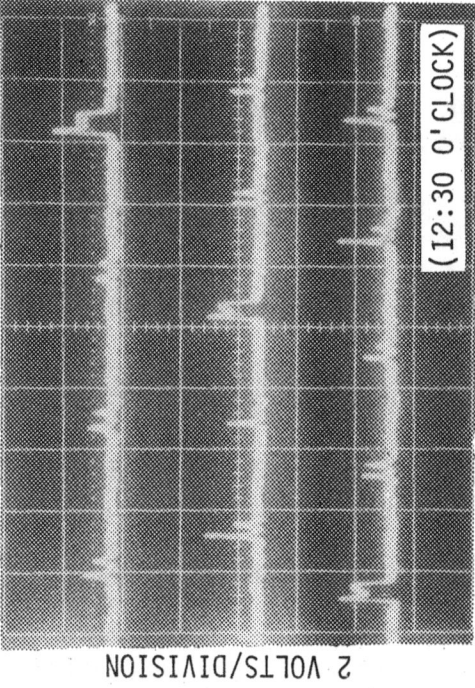

2 VOLTS/DIVISION

(12:30 O'CLOCK)

5 MILLISECONDS/DIVISION

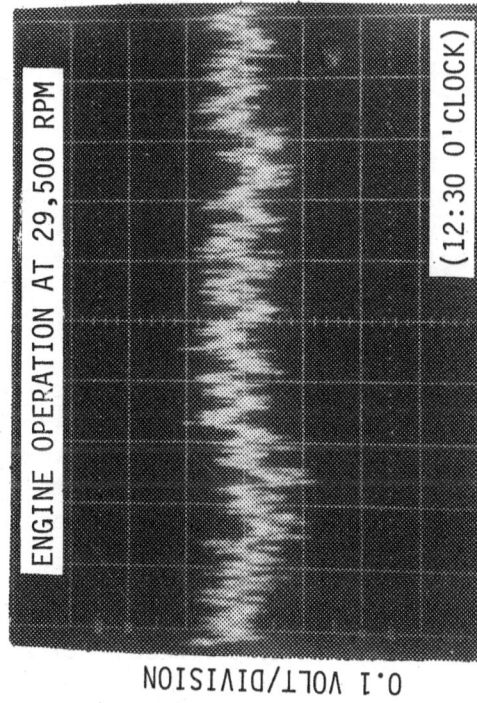

ENGINE OPERATION AT 29,500 RPM

0.1 VOLT/DIVISION

(12:30 O'CLOCK)

1 MICROSECOND/DIVISION

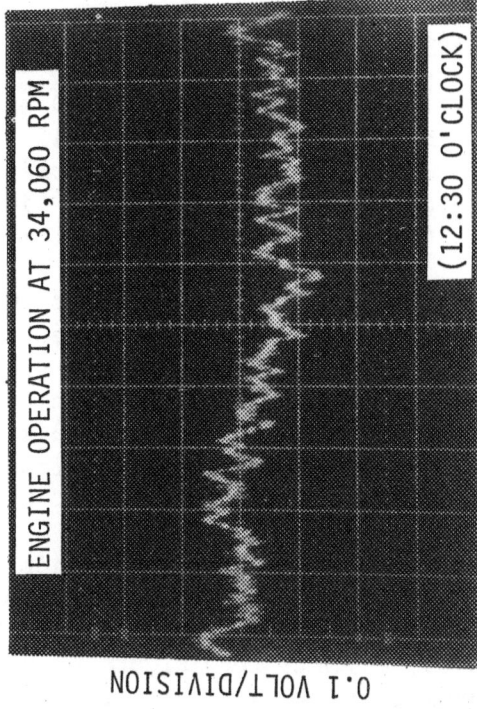

ENGINE OPERATION AT 34,060 RPM

0.1 VOLT/DIVISION

(12:30 O'CLOCK)

0.1 MICROSECOND/DIVISION

Figure 44.   System Output For Engine Manual Rotation and Engine Operation at 29,000 and 34,060 RPM.

89

Further review of the data and systems indicated that during the motoring, electrical interference was evident. Additionally, damage was noted in one fiber optic bundle.

Engine Test No. 3

Two systems were installed in 6:30 and 12:30 o'clock positions. The touch probe was installed in the 12:00 o'clock position. For this test, the optical filter was not incorporated into the system in order to avoid the attenuation loss of the laser light for the lower power conditions, for which the emitted radiation from the blade would be low.

Data from the MTCMD was recorded on magnetic tape and on film from oscilloscopes. Touch probe data was recorded manually from a digital readout. The system at 6:30 o'clock showed signals at initial rollovers and idle indicative of clearance measurements. Signals from the 12:30 o'clock system were of low amplitude, insufficient to provide clearance information. On the first acceleration from idle to higher speed, the 6:30 o'clock system ceased functioning.

Data was recorded at rollover speeds of 11,000 and 20,000 rpm and at idle, 29,000 rpm. The 6:30 o'clock system generated distinct clearance signals from the solid tip blades as shown in the on-line photographs, Figures 45, 46, and 47. Figure 48 shows similar data from tape playback. The system at 12:30 o'clock gave signals which appeared to be generated by passing blades, but the signal amplitude was insufficient to provide clearance data - apparently due to the damaged input fiber optic bundle. This system also gave signals which appear to result from passing blades at IRP following several burst and chops between idle and IRP. It is anticipated that these signals were due to infrared emission from the blades through the undamaged fiber optic output bundle, rather than from the laser reflection. Signals from the 12:30 o'clock system are shown in Figure 49.

During this run, a failure in the cooling system supplying CO_2 to the probes resulted in damage to them. Prior to failure, the cooling system was effective in keeping the temperature of the probe head below 90°F. Several operational amplifiers in the electronics package also failed despite forced air cooling.

Changes Made After First Three Tests

Following the first three engine tests, several changes were made to the clearance measurement system. A major improvement was made to the laser-fiber optic interface and refinements were made in the electronic amplifier and divider package.

It was decided to use polished CODEP coating on the blade tips in order to raise the level of reflectivity and reduce oxidation during test.

ON-LINE OSCILLOSCOPE MULTISWEEP PHOTOGRAPHS

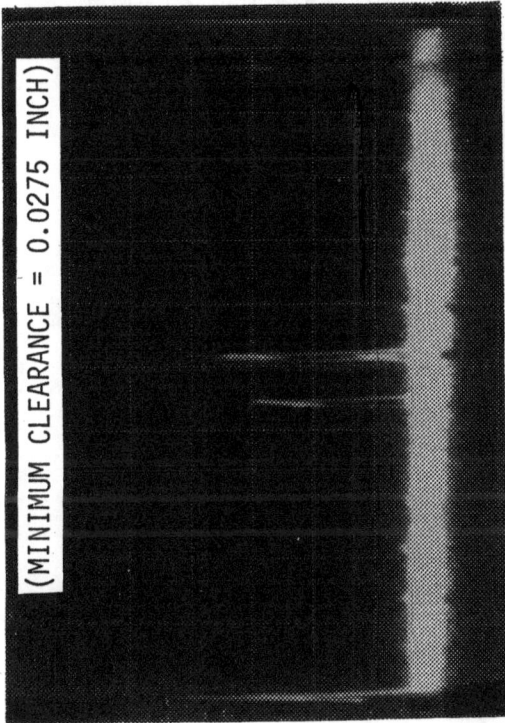

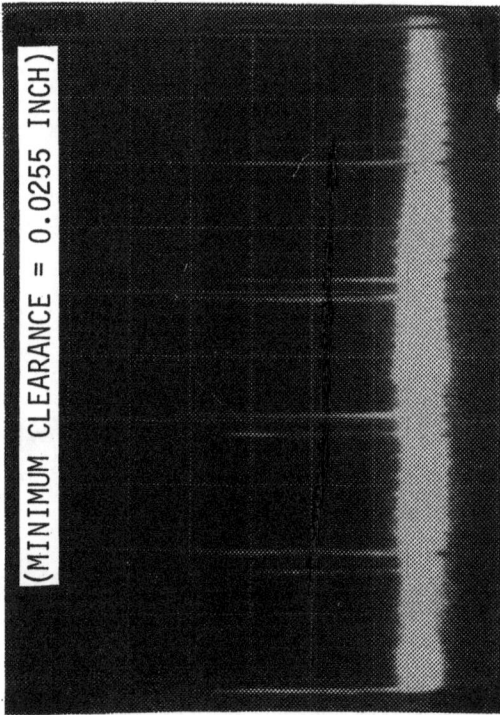

(MINIMUM CLEARANCE = 0.0255 INCH)

(MINIMUM CLEARANCE = 0.0275 INCH)

ON-LINE OSCILLOSCOPE SINGLE SWEEP PHOTOGRAPH

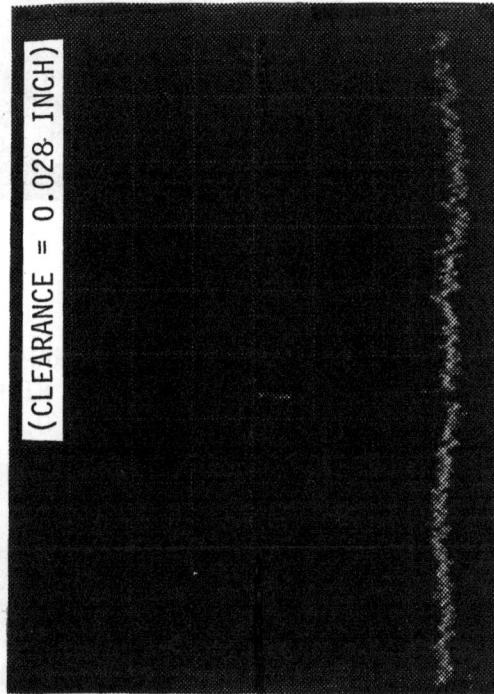

(CLEARANCE = 0.028 INCH)

Figure 45. Probe Readings Taken at 6:30 O'Clock for Engine Rollover - Approximately 10,000 RPM.

91

ON-LINE OSCILLOSCOPE PHOTOGRAPH AT 11,000 RPM, SHOWING APPARENT
CLEARANCES OF (FROM LEFT TO RIGHT): BLADE NO. 8 = 0.0235 INCH,
NO. 15 = 0.0265 INCH, NO. 22 = NO DATA, NO. 29 = 0.028 INCH, AND
NO. 1 = 0.031 INCH.

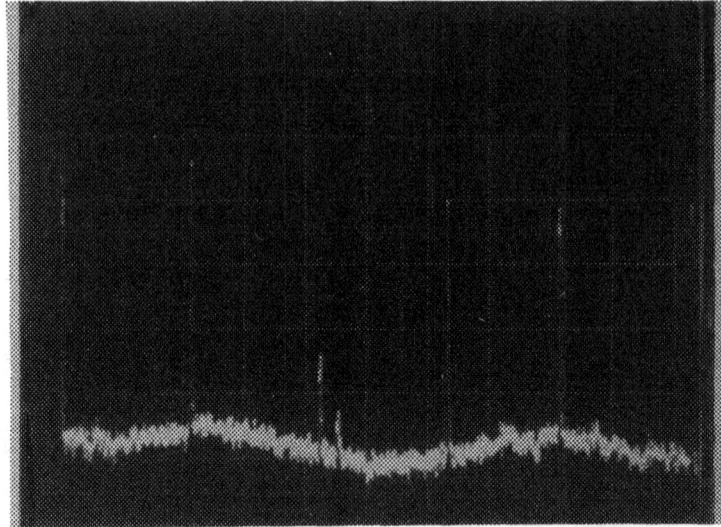

ON-LINE OSCILLOSCOPE PHOTOGRAPH AT 20,000 RPM, SHOWING APPARENT
CLEARANCES OF (FROM LEFT TO RIGHT): BLADE NO. 29 = 0.0325 INCH,
NO. 1 = 0.037 INCH, NO. 8 = 0.035 INCH, AND NO. 15 = 0.034 INCH.

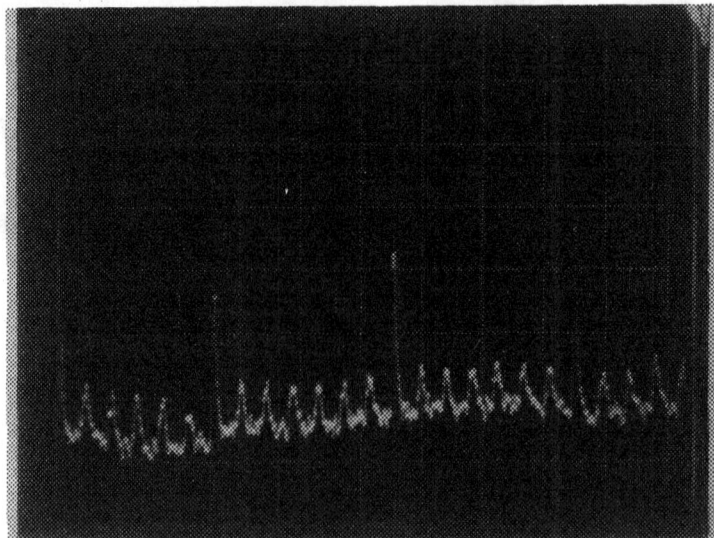

Figure 46.   Probe Readings at 6:30 O'Clock - 11,000 and 20,000 RPM.

ON-LINE OSCILLOSCOPE PHOTOGRAPH AT 29,000 RPM (IDLE),SHOWING APPARENT
CLEARANCES OF (FROM LEFT TO RIGHT): BLADE NO. 1 = 0.0435 INCH,
NO. 8 = 0.0405 INCH, NO. 15 = 0.0405 INCH, NO. 22 = 0.046 INCH.

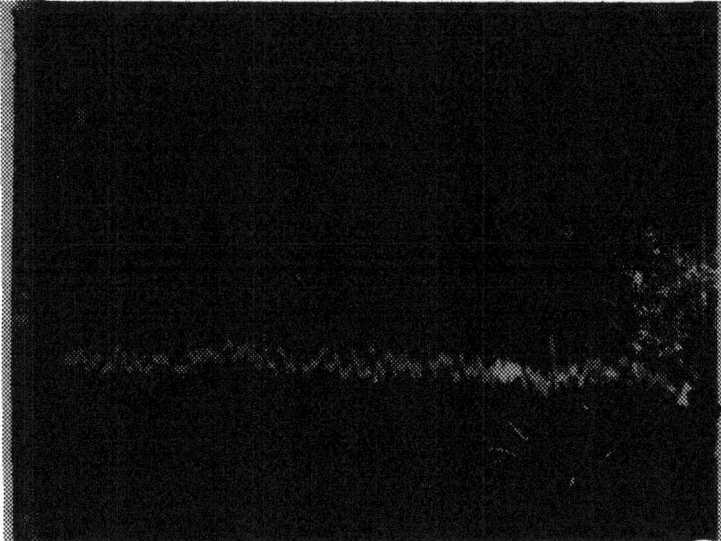

ON-LINE OSCILLOSCOPE PHOTOGRAPH AT 29,000 RPM (IDLE),SHOWING APPARENT
CLEARANCES OF (FROM LEFT TO RIGHT): BLADE NO. 1 = 0.045 INCH,
NO. 8 = 0.042 INCH, NO. 15 = 0.0435 INCH, NO. 22 = 0.048 INCH, AND
NO. 29 = 0.040 INCH.

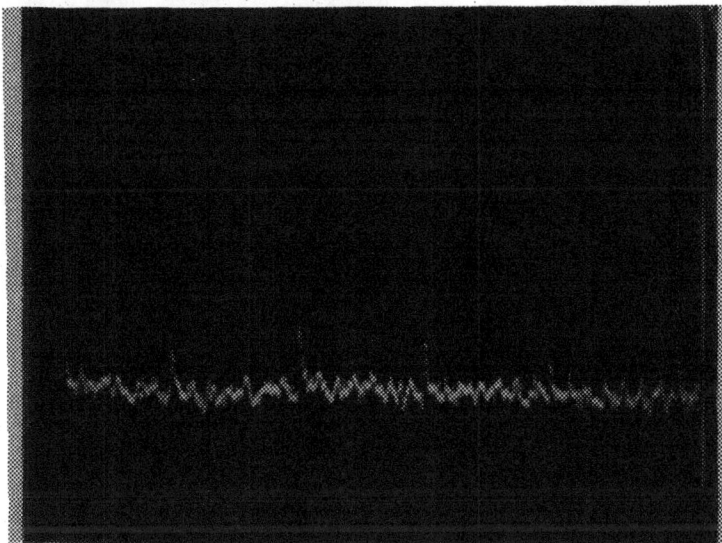

Upper photograph was taken early in the 5-minute run,while the lower
photograph was taken two to three minutes later in the same run. Note
the deterioration in the signal level.

Figure 47. Probe Readings at 6:30 O'Clock - 29,000 RPM.

93

IDLE

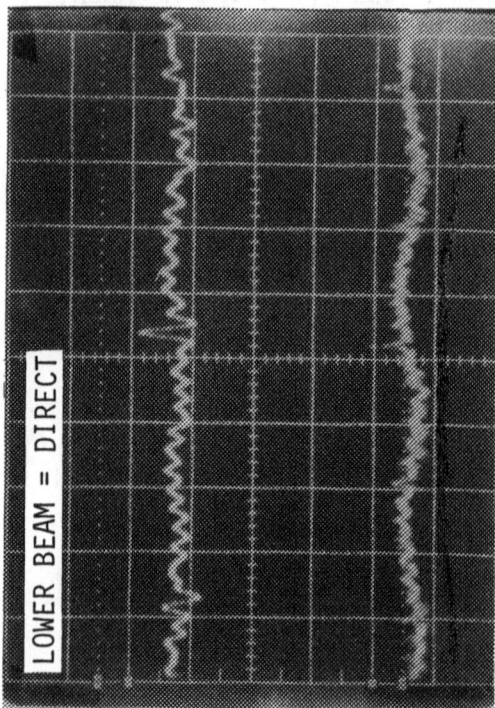

LOWER BEAM = DIRECT

ROLLOVER

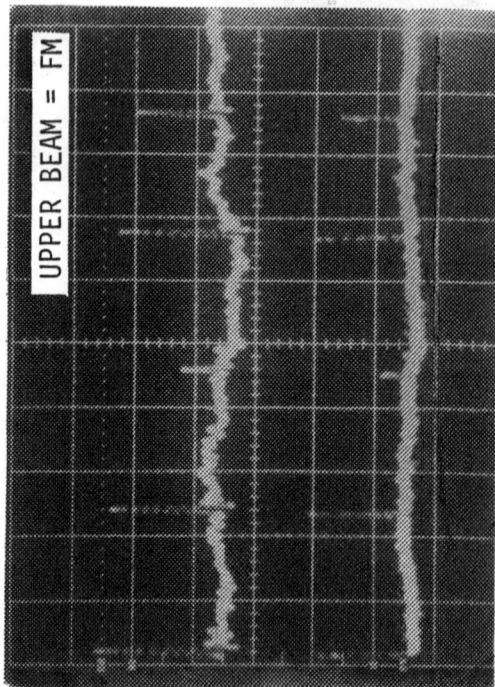

UPPER BEAM = FM

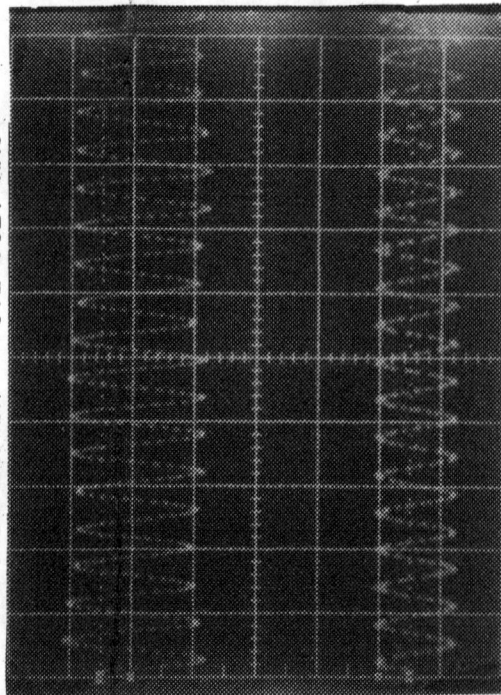

CALIBRATION = 0.2 VOLT RMS

Figure 48. 6:30 O'Clock Magnetic Tape Playback of FM and Direct Record.

94

ON-LINE OSCILLOSCOPE PHOTOGRAPH AT ROLLOVER (10,000 RPM)

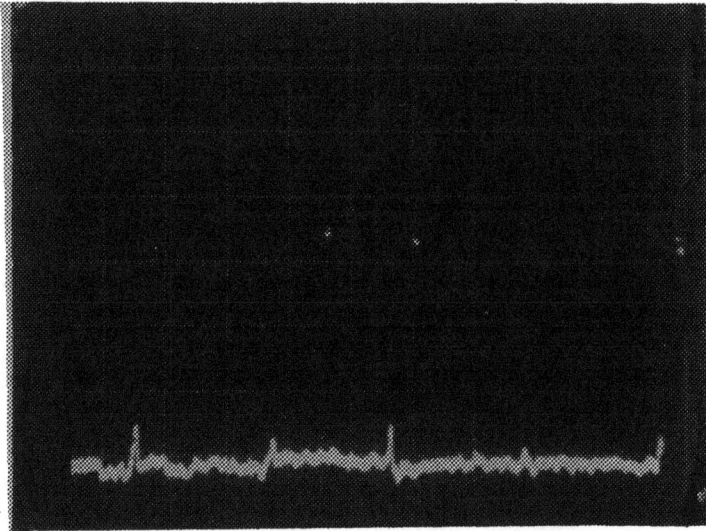

ON-LINE OSCILLOSCOPE PHOTOGRAPH AT IDLE (29,000 RPM)

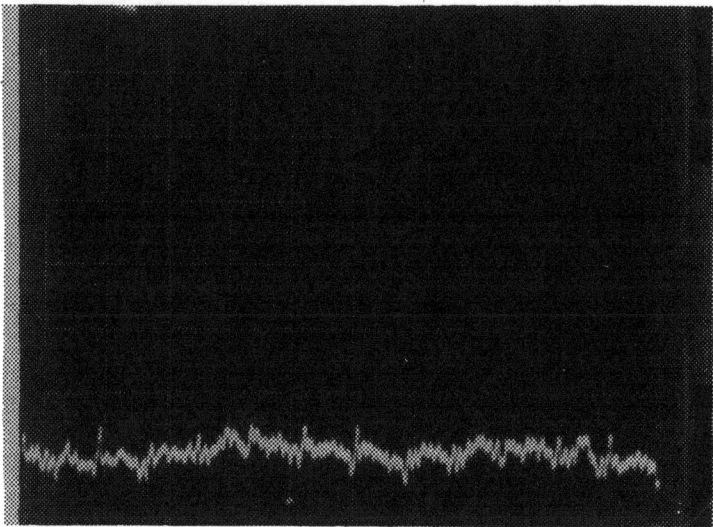

The above data was suspected to be incorrect because of apparent damage to the input fiber optic bundle. The low levels are indicative of the problem; however, the waveform still shows the solid tip blades.

Figure 49. Probe Readings at 12:30 O'Clock.

Probe Revisions: The method of clamping the input bundle to the laser and lens system was judged to be unsatisfactory. Set screws alone were inadequate for precision positioning of the fiber bundle to the focal point of the microscope lens within a tolerance of 0.00075 inch. A three-axis positioner with micrometer heads was constructed to locate the single fiber, masked with a precision pinhole, in the focal point of the microscope objective (Figure 50). Through careful choice of pinhole size, it was possible to visually set the bundle position while the probe was mounted in the engine. The lens was mounted rigidly to the laser, and the laser lens and 3-D positioner were mounted on a rigid watercooled plate for minimum differential expansion. A 0.001-inch pinhole was mounted over the end of the input bundle where it entered the probe. This prevented spurious illumination from light, which leaked from the active fiber to other fibers ("fiber crosstalk").

Metal sheathing was added to the fiber optic bundles for protection, and additional forced air cooling was added to the probe assembly and fiber bundles near the engine.

Refinements in Electronics/Detector Package: At this time, a measurement of the speed of response of the entire electronics system, including SC-10 diode and signal processing circuitry, was made. A light-emitting diode was driven with known voltage pulses at repetition rates from 1 Khz to 1 Mhz. System voltage output was steady over the low frequency range declining at a break point of 30 Khz and above at a 12-decibel per octave rate. Since clearance signals have significant energy content at 30 Khz and above while vibration frequencies are lower, it was decided to use a high-pass filter in the data recording system to counteract the preferential gain given to low-frequency signals by the signal processor and detector.

Further review of the system indicated that the loss of energy through the band-pass filter and other optical components of the probe coupled with low reflectivity of the blade tips required very high gain electronics. The need for high gain and extremely wide band width requires fast, stable amplifiers. The system electronic layout was changed to reduce electrical noise and instability. The gain in the amplifiers was increased up to the threshold of instability (approximately 10,000 to 1). Consideration was given to reverse biasing of the detector to increase gain and speed of response; however, this would be incompatible with system stability and would have required a significant development effort.

Installation Changes: Each laser was rigidly clamped on a water-cooled base so that there would be no motion relative to its illuminated fiber optic bundle. The laser was allowed to expand thermally only at the rear mounting. The returning (output) fiber optic bundle was clamped in a fixture to its detector and to the mounting plate. Each plate held an enclosure for heat and noise isolation, and each probe, including the touch probe, received forced air cooling of exposed components as well as the original cooling system.

Figure 50. Three-Axis Positioning System.

TV System: Prior to the fourth engine test, a TV (MTCMD) position detection system was constructed. The laboratory calibration setup is shown in Figure 51. This provided an additional check on probe operation and on clearance data.

The adapter between the TV camera and fiber optic output bundle was similar to the P&WA design. However, in addition to the more selective band-pass optical filter, this design included a microscope objective lens mounted in a special adapter, spacing it from the TV camera cathode at the proper distance for a greater magnification (160:1).

The TV system also included the improved mounting plate, isolation and cooling systems, as well as the precision alignment system developed in the laboratory. Various pinhole aperature stops were tried with this system. The smallest spot size obtained in laboratory tests was 0.008 inch at the output bundle. This was achieved by putting a 0.001 inch pinhole over a single optical fiber at the input where the entire laser output would be focused. Once this pinhole was installed, the same fiber was identified at the probe end of the input bundle. Another 0.001 inch pinhole was installed over the fiber. Thus, no other fiber could contribute to the illumination of the probe by means of "cross talk".

However, it was determined that these apertures caused a substantial reduction in stray light (which was exaggerating the spot size). This stray light was nearly half of the optical power output of the probe without benefit of the pinholes, and the pinholes were changed to 0.003 inch diameter to increase energy to both the TV camera and the diode.

Another benefit of increasing the size of the pinholes to 0.003 inch was that adjustment of the input optical system was much easier since the focused laser beam had nine times the area of the 1 mil aperture in which to fall. It was possible to adjust both the TV system and the GE system in the test cell with probes installed on the engine, between engine runs. Reflected spot size for the TV system appeared as 0.008 inch with the 0.003-inch pinholes on the input fiber optic bundle as the TV moniter did not detect the stray light.

Calibration: During the course of calibrating the systems for engine test No. 4, a series of measurements was made in order to determine the reason for the poor repeatability of calibration curves. Since the analog divider circuit is very sensitive to its trim adjustments and to drift in the summing and differencing amplifiers when operating with small signals, it was decided to obtain an extended series of calibration curves, making each target traverse a number of times over an extended time interval without any electronic signal processing except for the current-to-voltage conversion amplifiers, A1 and A2. Voltage outputs of these amplifiers were recorded and later reduced by computer to form the ratio of the difference to the sum. Such an experiment would link repeatability of calibration (hence, the reliability of measurement) to the relationship between optical power and signal voltage.

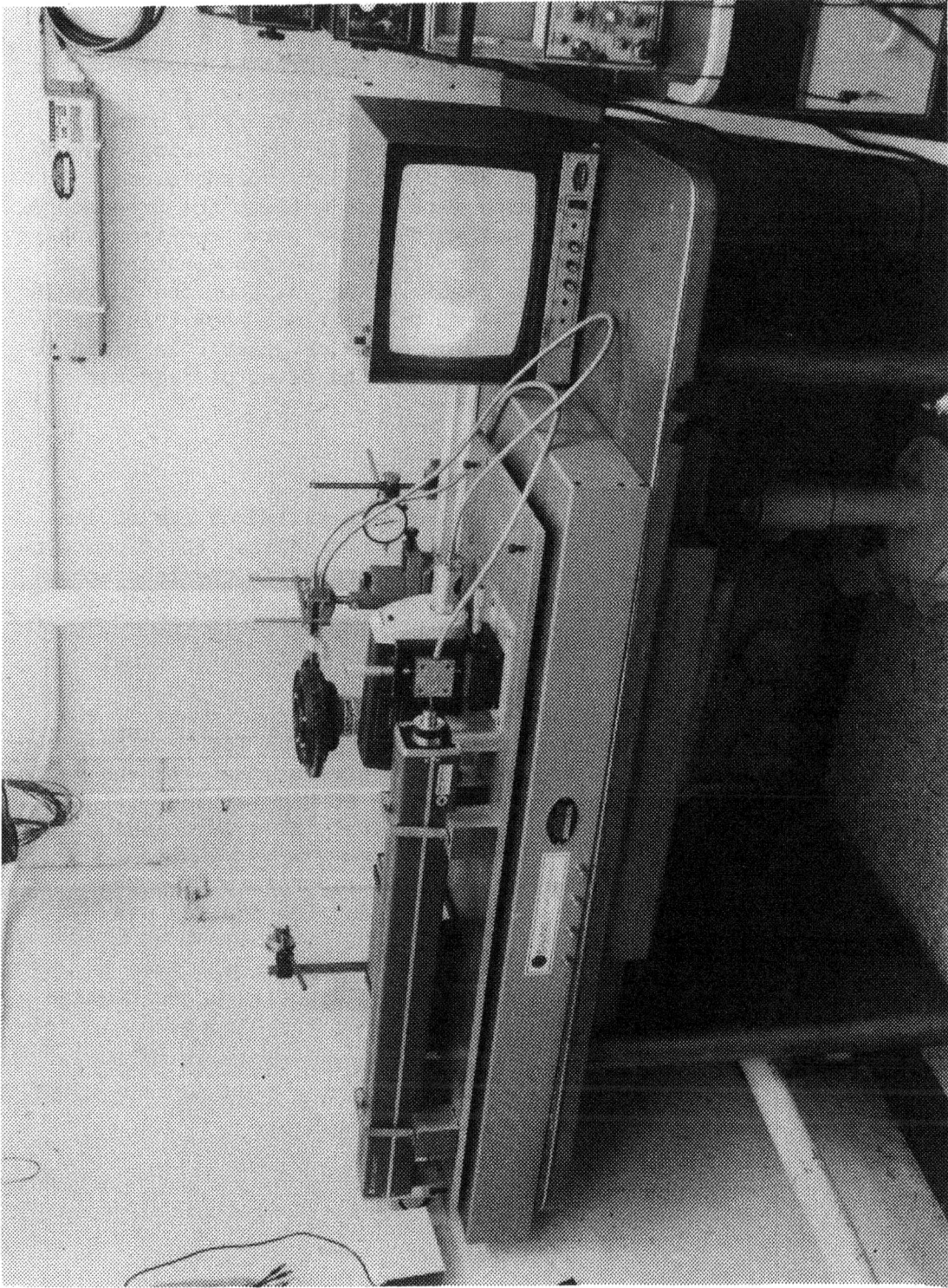

Figure 51. MTCMD TV System.

Figures 52 through 56 show that not only is the calibration sensitive to target reflectivity (all curves should ideally be the same), but there is a striking correlation between the spread in the curve and the optical power incident on the detector.

From these figures, it is seen that for a given calibration clearance, a range of current ratio values will be measured. Conversely, for a given current ratio, a range of clearances can occur. This range of possible clearances represents the uncertainty or accuracy of the detector system. Figure 57 shows the result of plotting the uncertainty versus diode optical power. Since these measurements were made without the presence of the analog divider, the hyperbolic appearance of the curve confirms that the effect of a fluctuation in a small quantity in the denominator of a division operation is magnified in the result. The shape of the curve in Figure 57 is an inverse power curve. These calibrations show that the optical power on the diode should be 25 microwatts or greater in order to obtain accurate clearance measurements and that the detector requires at least 10 microwatts to function effectively.

Engine Test No. 4

Both the TV and Electronic (diode) systems were installed on the engine and checked for operation by manual rollover of the engine, the diode system signal was noted to be weaker than when calibrated in the laboratory. Direct measurement of the signal in the output bundle revealed only 4 microwatts available to the detector versus 26 microwatts in the laboratory.

At idle speed, no usable signals were obtained from either system. The TV system functioned well at rollover but the signal was lost as the speed was raised. This was attributed to having only five blades with the solid tips. The infrequent passages of visible blades at high speeds did not supply a sufficient response time for the relatively slow TV scanning.

The systems were removed and recalibrated in the laboratory. Another attempt on the engine met with the same results and the testing was terminated.

Inspection of the engine revealed a reddish-brown deposit on the turbine blades which adversely affected the reflectivity of the blades.

MIRROR POLISHED STAINLESS STEEL TARGET

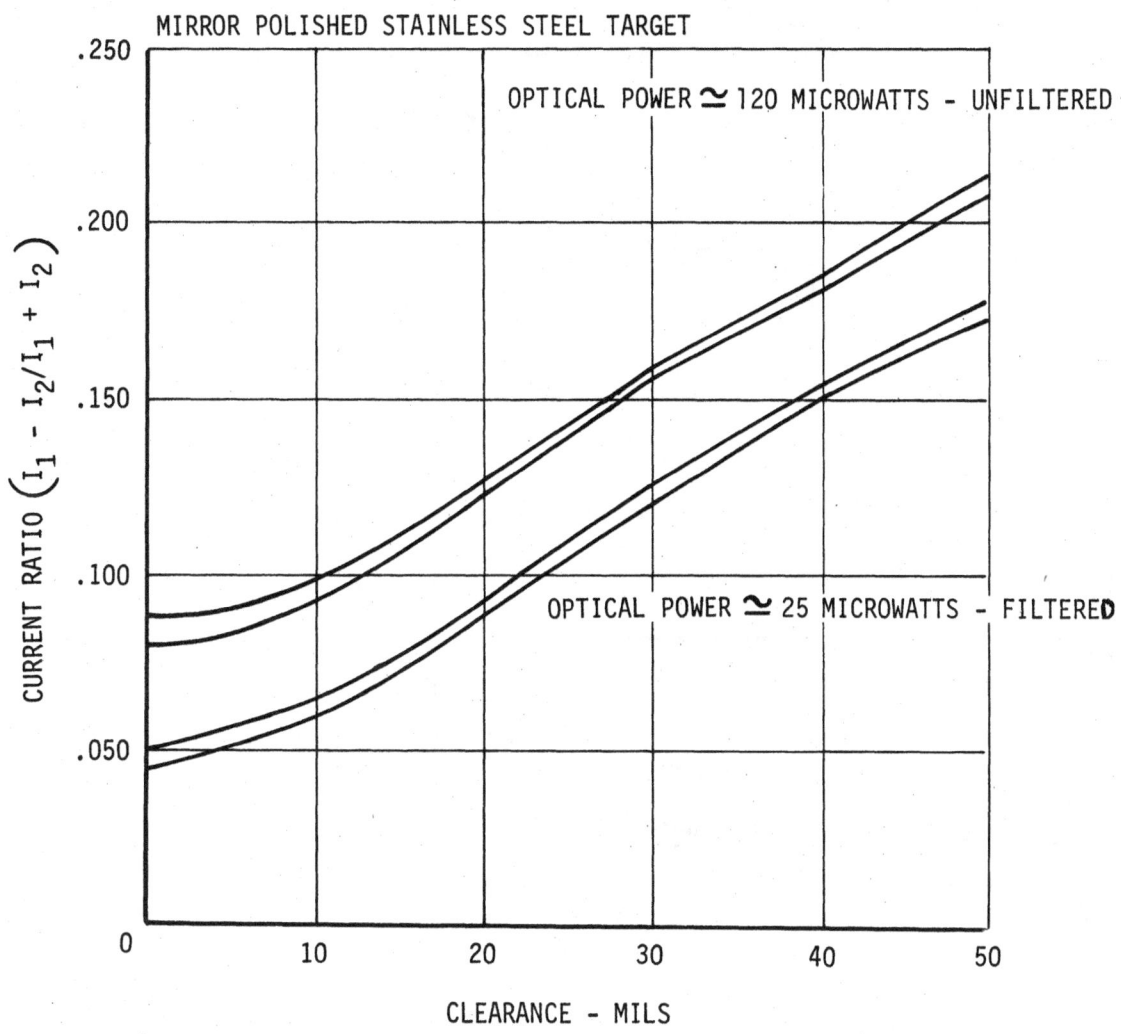

Figure 52. Calibration With Reflective Target.

101

DULL STAINLESS STEEL TARGET

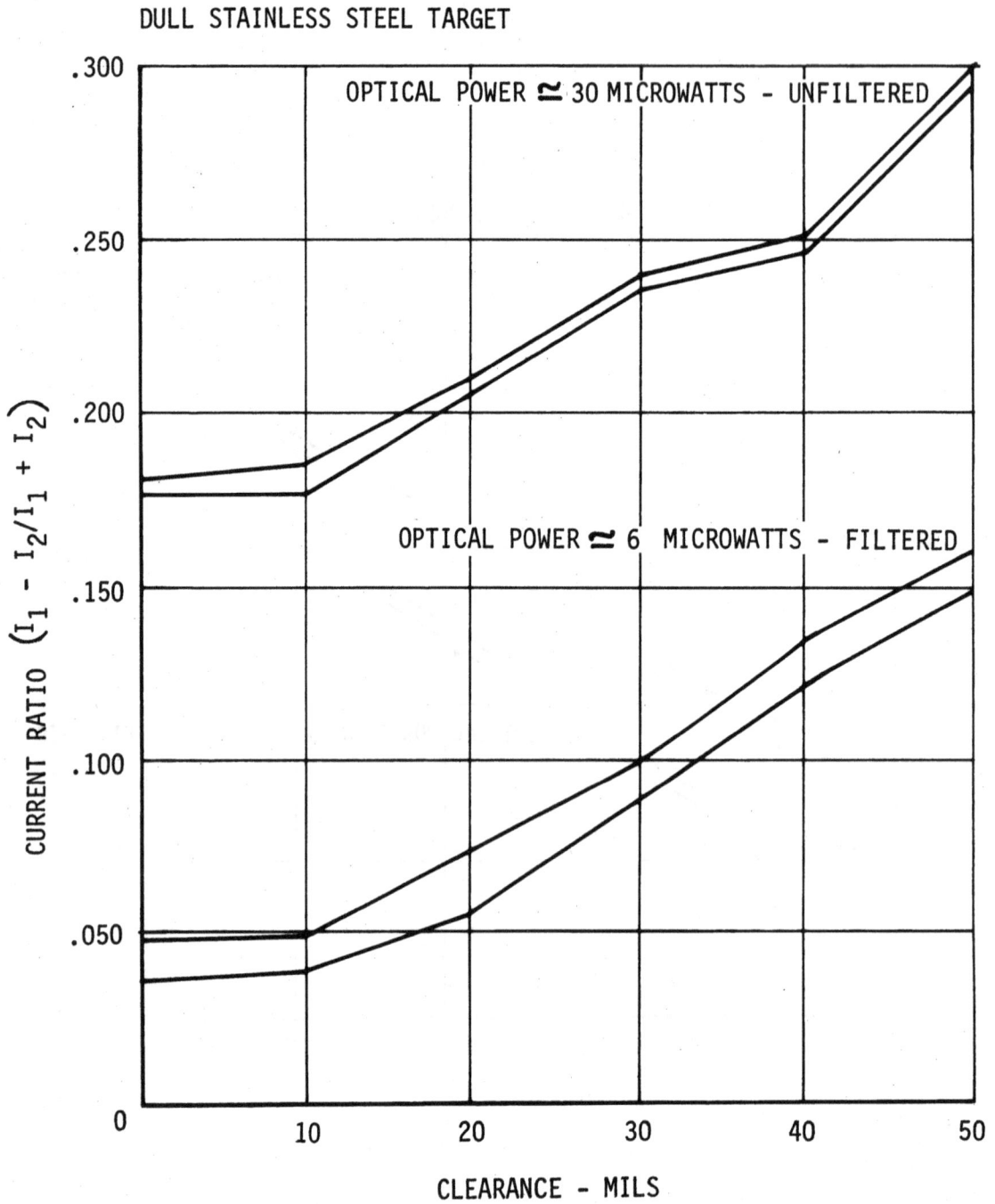

Figure 53. Calibration With Dull Target.

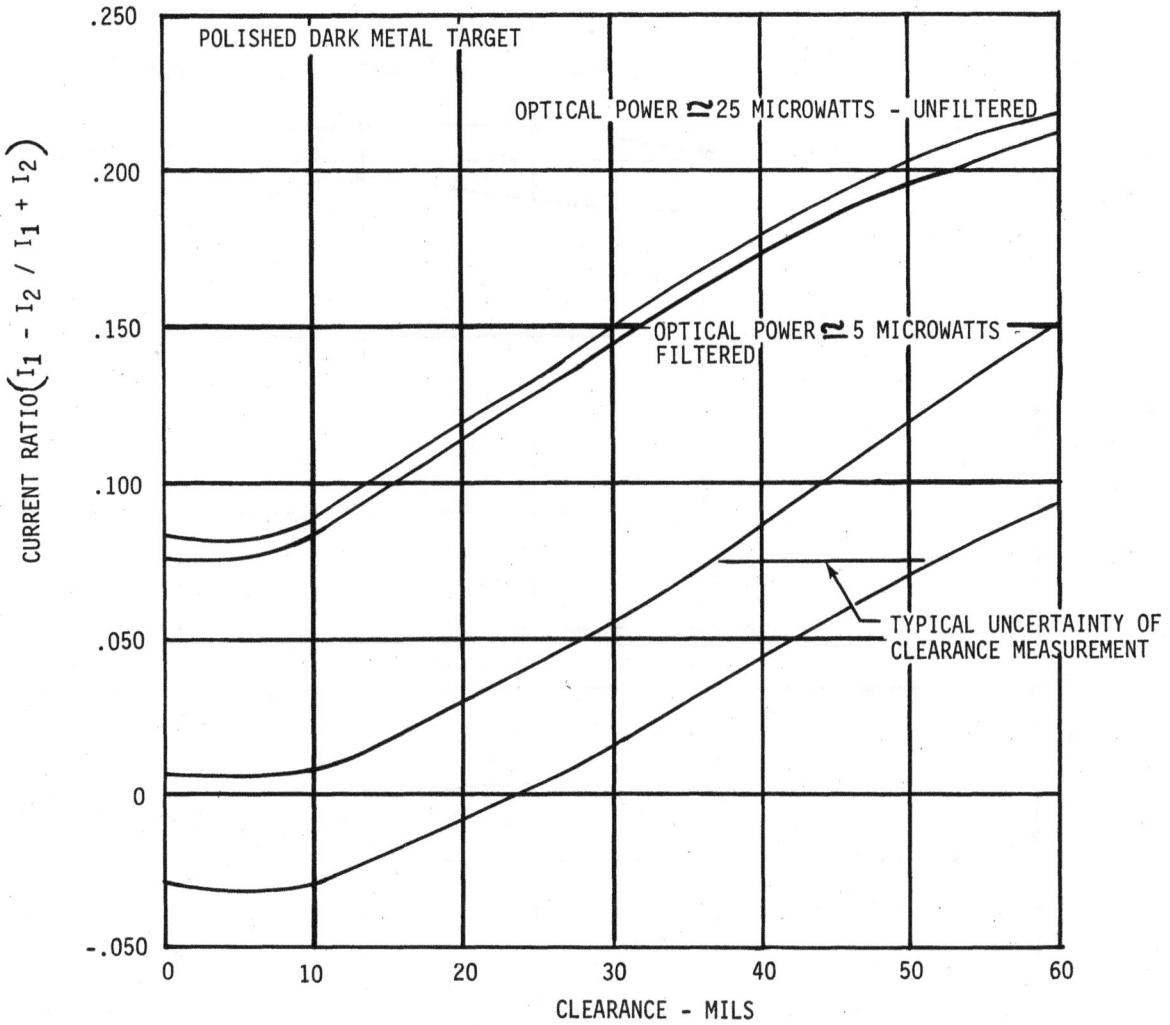

Figure 54.   Calibration With Dull Target.

103

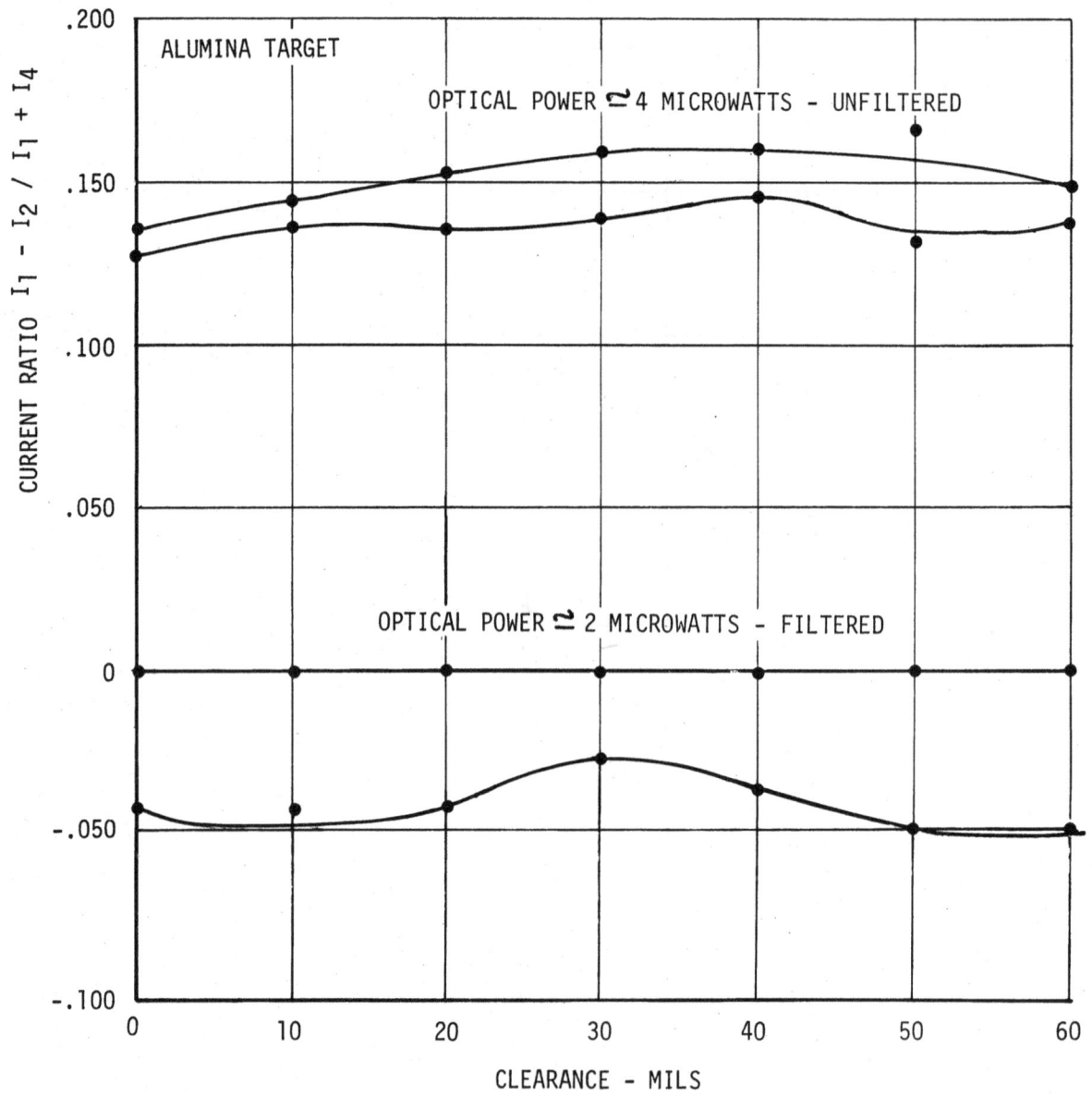

Figure 55. Calibration With Nonreflective Target.

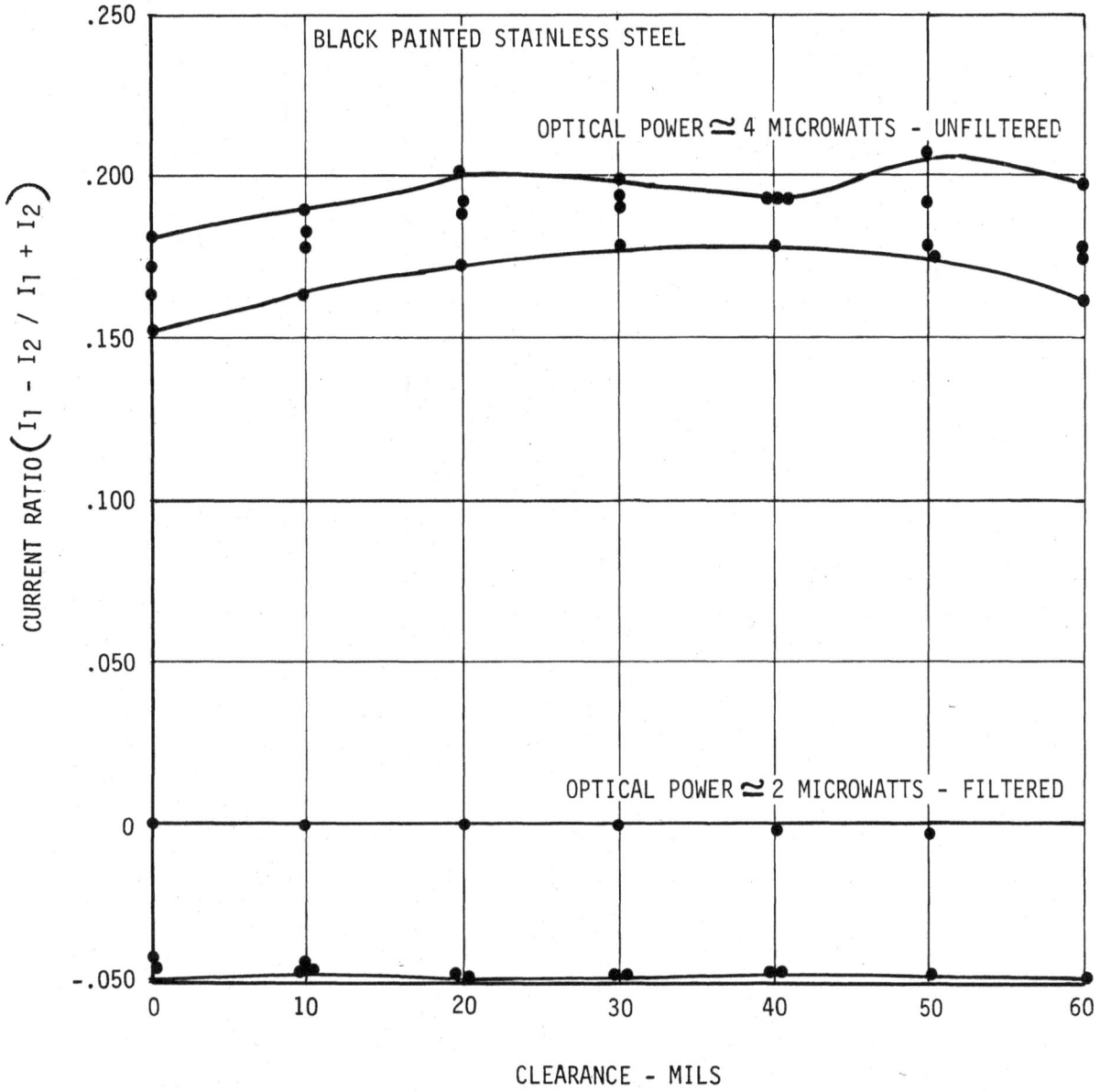

Figure 56. Calibration With Nonreflective Target.

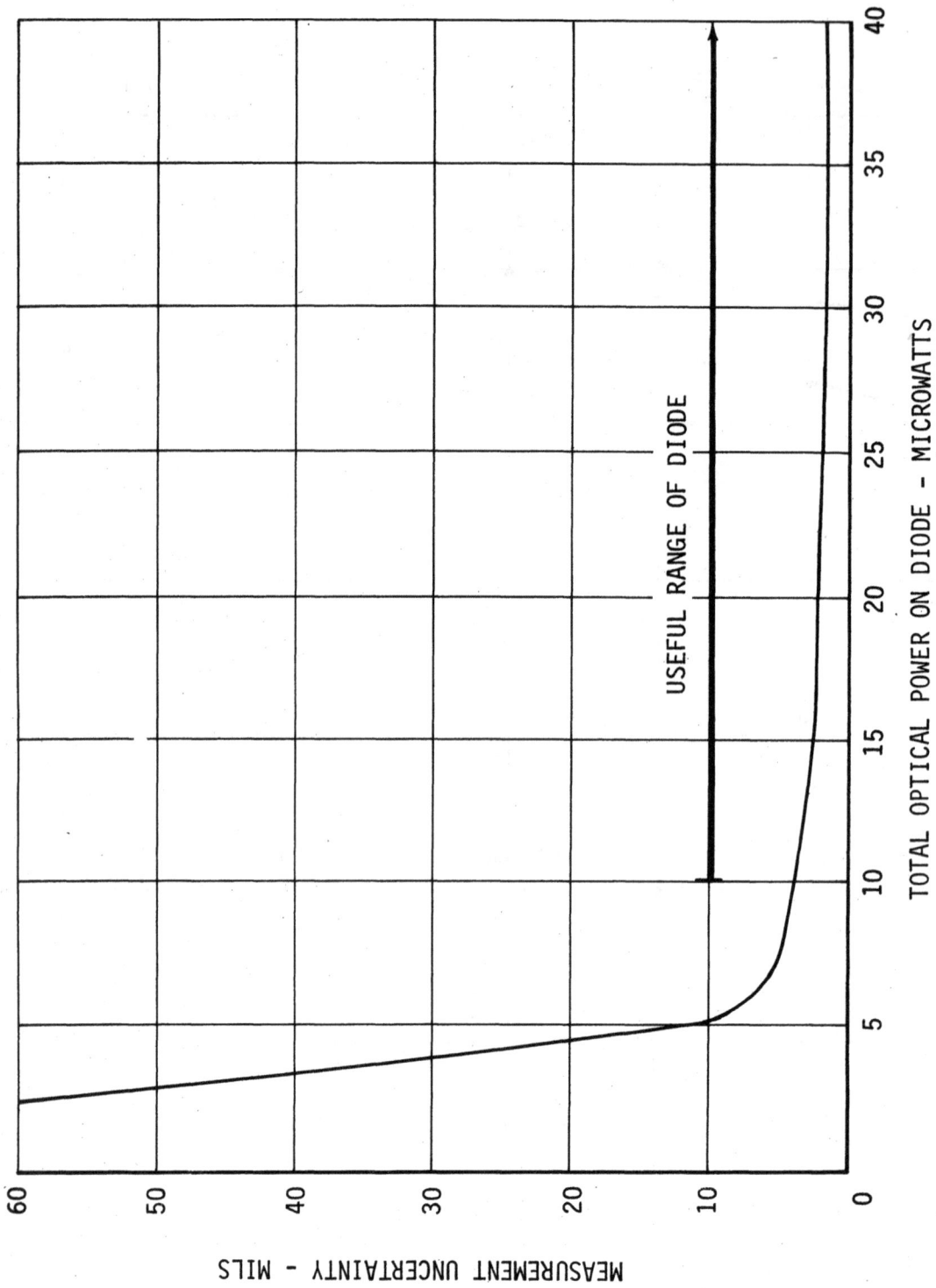

Figure 57. Clearance Uncertainty for Various Target Outputs.

The graph shows MEASUREMENT UNCERTAINTY - MILS on the vertical axis (0 to 60) and TOTAL OPTICAL POWER ON DIODE - MICROWATTS on the horizontal axis (0 to 40). The curve declines steeply from high uncertainty at low optical power, leveling off around 5 microwatts. An arrow labeled "USEFUL RANGE OF DIODE" spans from approximately 10 to 40 microwatts.

RECOMMENDATIONS FOR MTCMD SYSTEM IMPROVEMENTS

Introduction

The environmental conditions under which the clearance measurement system was tested, both in the laboratory and on the engine, revealed a number of problems which should be considered relative to additional development of the system.

The suggested solutions to these problems are complex and interrelated, indicating need for a fundamental approach involving:

1. System design optimization.

2. Extensive component and system bench tests with simulated combined environmental evaluation.

3. Development of the system for accuracy, durability, and ease of installation.

4. Verification of the system capability on an engine test

Of particular importance is a strong laboratory program that establishes full capability of the system as an accurate and reliable device prior to its operation on an engine.

Design Considerations

The following summarizes the significant problem areas and their resultant effect on the system operation.

PROBLEM AREA	EFFECT ON SYSTEM
Turbine Radiation	Masks the laser beam signal at high turbine temperatures. Band-pass filter attenuates working signal.
Blade Tip Reflectance	Not constant, decreases rapidly at turbine temperature and decreases signal strength.
Blade Tip Shape	Squealer tip blades at high RPM's do not present a wide enough surface to enable detector to respond to signal.

PROBLEM AREA	EFFECT ON SYSTEM
Installation Vibration	Misaligns optics, moves fiber bundle with loss of signal or calibration. Causes electronic noise.
Electrical Interference	Electronic package in engine test cell is affected by other electrical devices.
Fiber Optic Bundle Fragility	Loss of signal, increased noise.
Laser Spot Size	Large spot reduces intensity, causes detector inaccuracy.
Optical System Efficiency	Level of power at system output insufficient to trigger detector.
Optical System Reflections	Leads to high background noise competing with signal.
Environment Temperatures	Engine radiation affects external parts causing failure, electronic drift, or misalignment at connection points.
Position Detector Calibration	Low input power causes inaccuracy or loss of signal.
Electronics Calibration	Low signals require high gain and loss of stability.
Continuous Laser Beam	Reflections between blade tip passings mask clearance signals.

Recommended Design Changes

Turbine Radiation: The infrared radiation from the blades competes with the red laser. A band-pass filter effectively screens all but the laser wavelength. The attenuation of the filter can be reduced by a lens in front of the filter to redirect the rays normal to the filter (Item A, Figure 58). An alternate approach to this problem is to use a green laser (Argon) light (Item P).

Blade Tip Reflectance: For general use, the system should function with blade tips which have realistic oxidation levels and corresponding reflectivity. The power input versus output should recognize these realistic requirements when setting laser power requirement (Item P). For experimental use, a polished coating on the tip could be developed. CODEP was not confirmed as a suitable coating on this program.

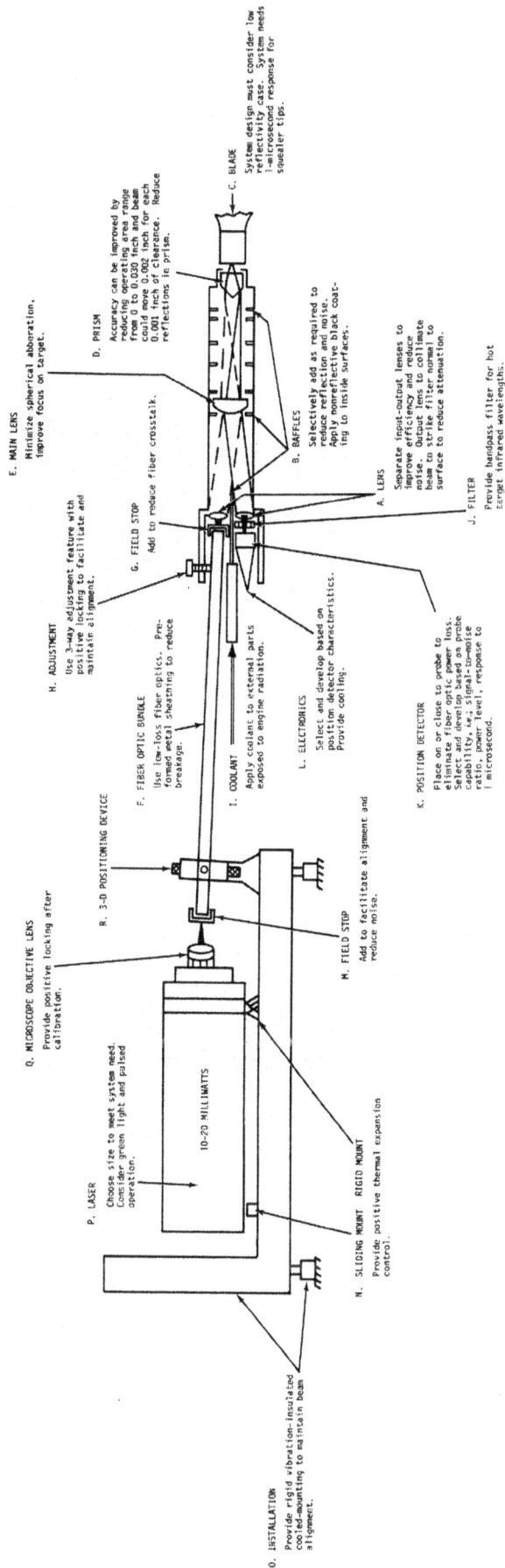

O. MICROSCOPE OBJECTIVE LENS
Provide positive locking after calibration.

R. 3-D POSITIONING DEVICE

F. FIBER OPTIC BUNDLE
Use low-loss fiber optics. Pre-formed metal sheathing to reduce breakage.

H. ADJUSTMENT
Use 3-way adjustment feature with positive locking to facilitate and maintain alignment.

E. MAIN LENS
Minimize spherical aberration, improve focus on target.

D. PRISM
Accuracy can be improved by reducing operating area range from 0 to 0.030 inch and beam could move 0.002 inch for each 0.001 inch of clearance. Reduce reflections in prism.

C. BLADE
System design must consider low reflectivity case. System needs 1-microsecond response for squealer tips.

G. FIELD STOP
Add to reduce fiber crosstalk.

B. BAFFLES
Selectively add as required to reduce reflection and noise. Apply nonreflective black coating to inside surfaces.

A. LENS
Separate input-output lenses to improve efficiency and reduce noise. Output lens to collimate beam to strike filter normal to surface to reduce attenuation.

J. FILTER
Provide bandpass filter for hot target infrared wavelengths.

I. COOLANT
Apply coolant to external parts exposed to engine radiation.

L. ELECTRONICS
Select and develop based on position detector characteristics. Provide cooling.

K. POSITION DETECTOR
Place at or close to probe to eliminate fiber optic power loss. Select and develop based on probe capability, i.e., signal-to-noise ratio, power level, response to 1 microsecond.

P. LASER
Choose size to meet system need. Consider green light and pulsed operation.

10-20 MILLIWATTS

M. FIELD STOP
Add to facilitate alignment and reduce noise.

N. SLIDING MOUNT   RIGID MOUNT
Provide positive thermal expansion control.

Q. INSTALLATION
Provide rigid vibration-insulated cooled-mounting to maintain beam alignment.

109

Blade Tip Shape: Blades with squealer tips (or indentations) will require a detector response as low as 1 microsecond to see the reflected signal of individual blades (Item K). Development of such a detector with matching electronics readout will probably be required to attain such a response. For experimental use, solid tip blades can be fabricated.

Detectors should not be limited to diode type during the system redesign study to take advantage of the latest advances in the various component elements.

Installation Vibration: In the test cell, engine vibration and movement of air can affect the fiber optic bundles and affect the signal. The fiber optic bundle should be housed in a rigid metal preformed sheath (Item F). The laser and electronics should be mounted in a rigid, vibration isolated enclosure.

Electrical Interference: Electronic devices utilized for normal engine test cell running create interference which affects the electronics of the position detection system (Item L). Filtering and standard means of screening competing signals must be provided for those devices which cannot be shutoff during clearance measurement tests.

Fiber Optic Bundle Fragility: On several occasions the 0.001-inch fibers in the bundles broke in spite of restricted handling. The bundles should be encased in metal sheathing to alleviate this problem (Item F).

Consideration should also be given to using larger diameter fibers at the input side and using an output lens to focus the reflected spot to the desired diameter (Item A).

Elimination of the output fiber bundle can be achieved by mounting the detector directly to the probe (with proper cooling, etc.) (Item K).

Laser Spot Size: A large spot falling on the position detector could adversely affect its accuracy. Field stops of 0.003 inch diameter at the output side of the input fiber optic bundle (Item M) tend to reduce the spot size by minimizing cross talk of the fibers at the output end.

The main lens (Item E) should be revised to a shape which improves the focus at the target (aspherical). The focal point on the target should be set to correspond to a clearance near the minimum expected, since this is the region of greatest interest.

Optical System Efficiency: The considerable power loss through the optical system experienced during this program (down to 0.1 percent of the laser power) can be improved. The three-lens system (Items A and E) tested by P&WA[1] should be adopted. Analysis showed an efficiency of 50% versus 25% for the two-lens system. Such a design would also enhance the design to collimate the rays through the bandpass filter (see Turbine Radiation).

As an alternate, a four-lens system could be considered. This would advantageously completely separate the input and output signals but would pose additional packaging challenges (see Figure 59). The state of the art for fiber optics is continually improving; new low loss fibers are available and should be used in this application.

Optical System Reflections: Reflections occur throughout the probe, primarily at the lenses and the prism, and create a noise level against which the signal must compete. The three-lens (or four-lens) system would significantly reduce such reflections. During the development of the probe the use of a cutaway (half probe) should be used so that the reflections can be observed and baffles (Item B) added at critical points within the probe. Consideration to splitting the prism should also be included (Figure 59).

Environment Temperature: The cooling applied to the probe body proved to be satisfactory; however, radiation from the engine caused the cemented fiber optic bundle end connector to loosen. Also the electronic packages were affected by this radiation and their own internally generated heat. The electronics were mounted on a water-cooled frame and supplied with forced cooling. Water-cooled pipes were wrapped around the probe head to protect the exposed fiber optics. In a redesigned probe the cooling for the internal probe should also encompass the exposed parts adjacent to the engine (Item I).

Position Detector Calibration: It is anticipated that most of the position detector problems could be resolved by incorporating the items previously described to provide a higher signal-to-noise ratio. Since the detector system is critical to the operation of the overall system, selection of the detector should be made on the basis of a complete system analysis. An assessment of the power loss through each component should be made during the design and development tests, and the selection of a detector should await assurance that a sufficient level of energy with the necessary characteristics is available.

During the system study, consideration should be given to the selection or development of position detectors that may operate on somewhat different principles from the one applied in this program.

Electronics Calibration: Adequate cooling, shielding and cabling together with a sufficiently high signal level are required to assure stable and accurate performance.

Continuous Laser Beam: The problem of noise due to reflections within the probe body (see Optical System Reflections) may be reduced by utilizing a pulsed laser beam synchronized to blade tip passings. The noise level within the probe would be present for a shorter duration; this would improve the performance of systems which use an integrating position detector or a TV output.

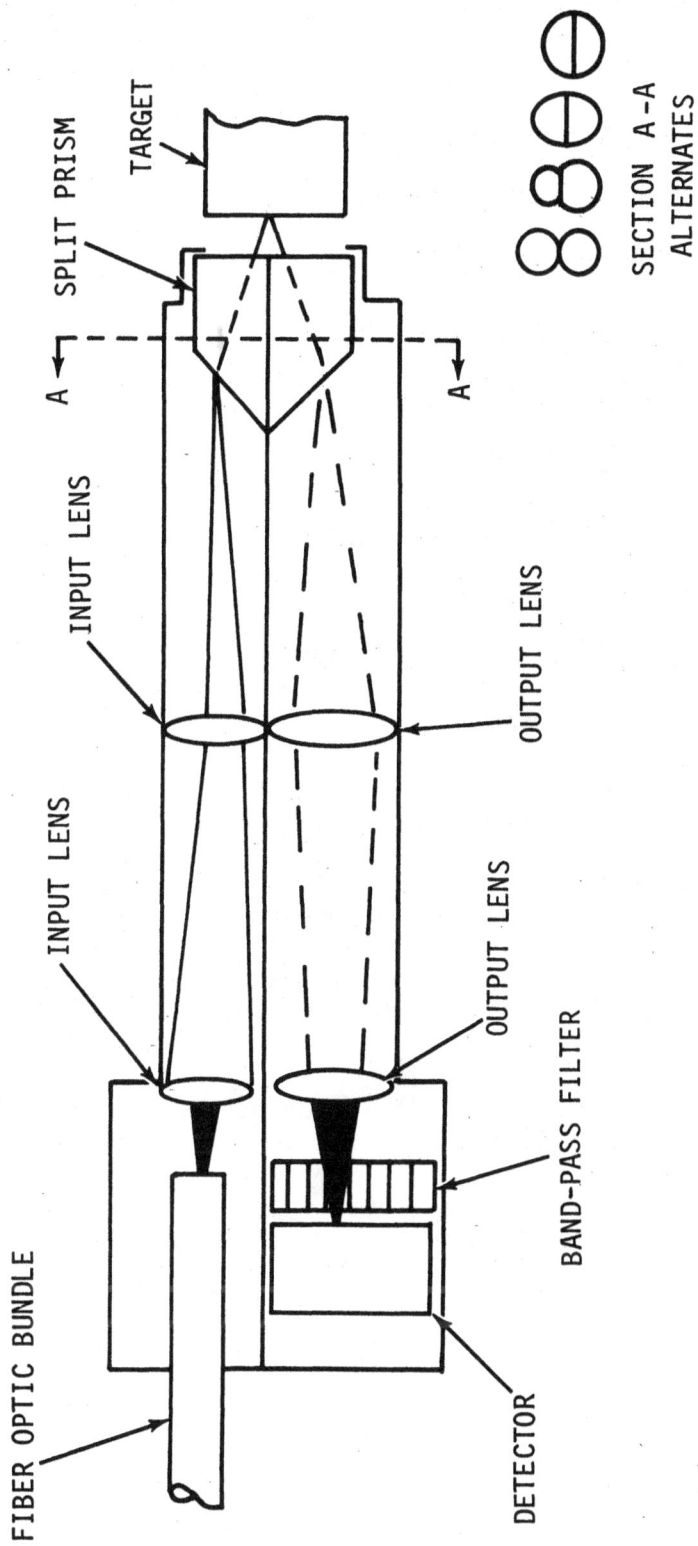

TARGET

SPLIT PRISM

INPUT LENS

INPUT LENS

INPUT LENS

OUTPUT LENS

OUTPUT LENS

OUTPUT LENS

BAND-PASS FILTER

DETECTOR

FIBER OPTIC BUNDLE

SECTION A-A
ALTERNATES

Figure 59. Four-Lens System.

CONCLUSIONS

The feasibility of the principles of a laser clearance measurement device utilizing a position detector and electric analog output has been demonstrated. The experience gained during this program indicates that significant development is required in the following areas:

- Target radiation.

- Blade tip reflectance.

- Blade tip configuration.

- Installation vibration.

- Electrical interference.

- Environment temperatures.

- Fiber optic bundle fragility.

- Laser spot size.

- Optical system efficiency.

- Optical system noise.

The analytical techniques utilized in calculating turbine tip clearances are in good agreement with measured clearances.

Taking the more refined analysis and the touch probe data as the references, the following conclusions can be made:

- Rotor and stator growth during transients continues over a longer time period than previous, less refined analysis indicated.

- Clearances above idle speed are from 2 to 5 mils smaller than predicted by the previous less refined analysis.

- Minimum clearance occurs approximately 10 seconds after initiation of a burst from idle to IRP.

114

RECOMMENDATIONS

Continue the development of a laser tip clearance measurement device which incorporates the features of General Electric's basic approach, i.e. an electric output (as opposed to a solely visual display) which permits:

- Processing for computation or control use.

- Convenient storage and retrieval of data.

- Individual blade clearance determination.

Conduct system design studies, component designs, and a thorough laboratory evaluation and development program for engine environmental and operating conditions to produce a universal system, readily adaptable to a range of engine configurations. Particular engine related areas that should be addressed include:

- Hot target radiation.

- Blade tip reflectivity.

- Response rate to varied blade tip configurations.

- Engine case radiation to probe system components.

- Electrical and vibration interference effects on system accuracy.

Utilize the analytical techniques as applied in this program together with touch probe measurements for design and development of new turbines and to assess clearance levels on current engines, until fully developed laser probes are available.

APPENDIX A

TEMPERATURE RESPONSE CHARACTERISTICS

The temperature response characteristics are shown in Figures Al through
A61 for each of the steady-state and transient engine conditions studied.
The key for the turbine rotor temperature nodes is shown in the nodal
model of Figure 4 and those of the turbine stator are shown in Figure 5.

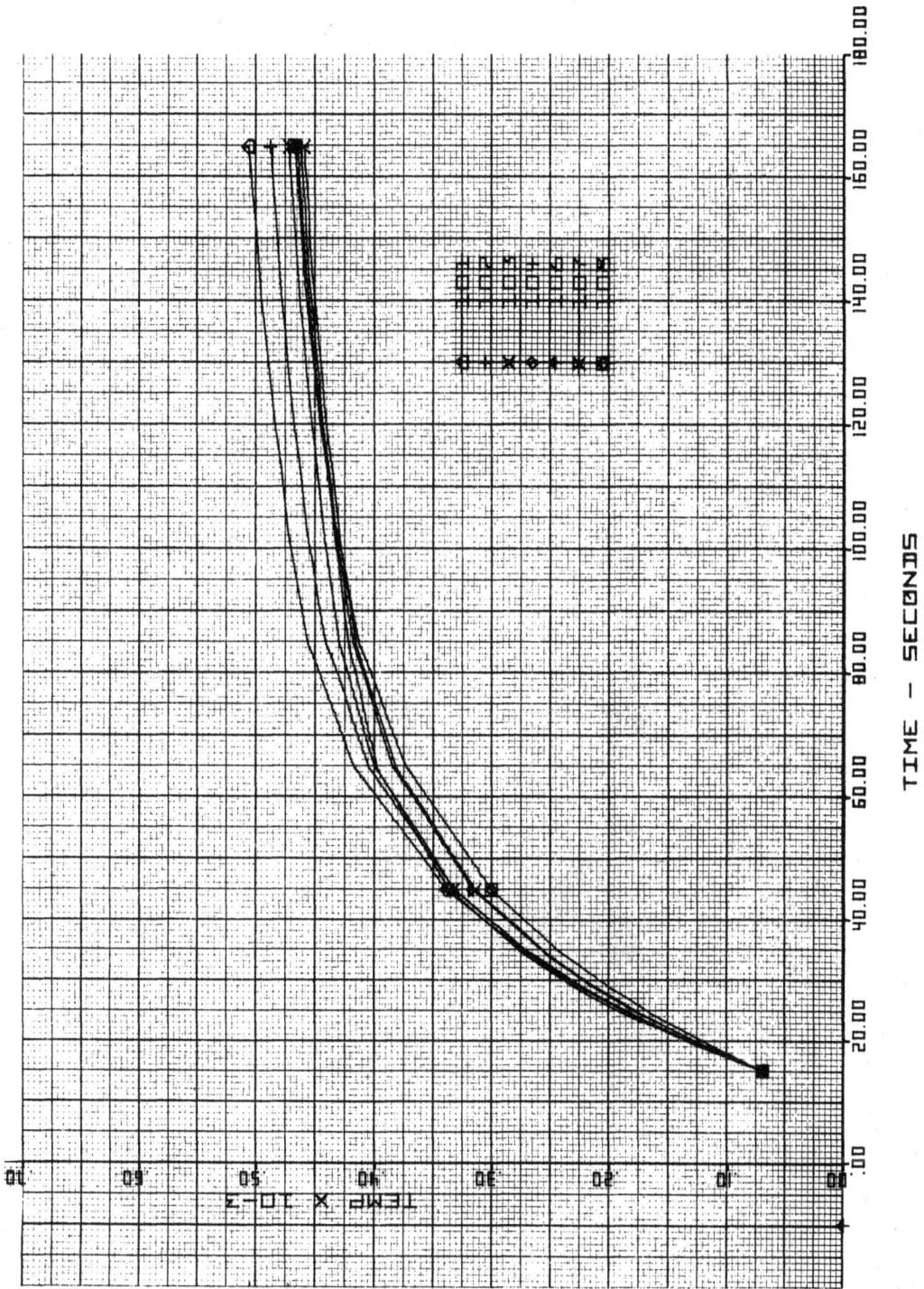

TIME — SECONDS

TEMP X 10-2

Figure A1. Forward Cooling Plate – Cold Start.

117

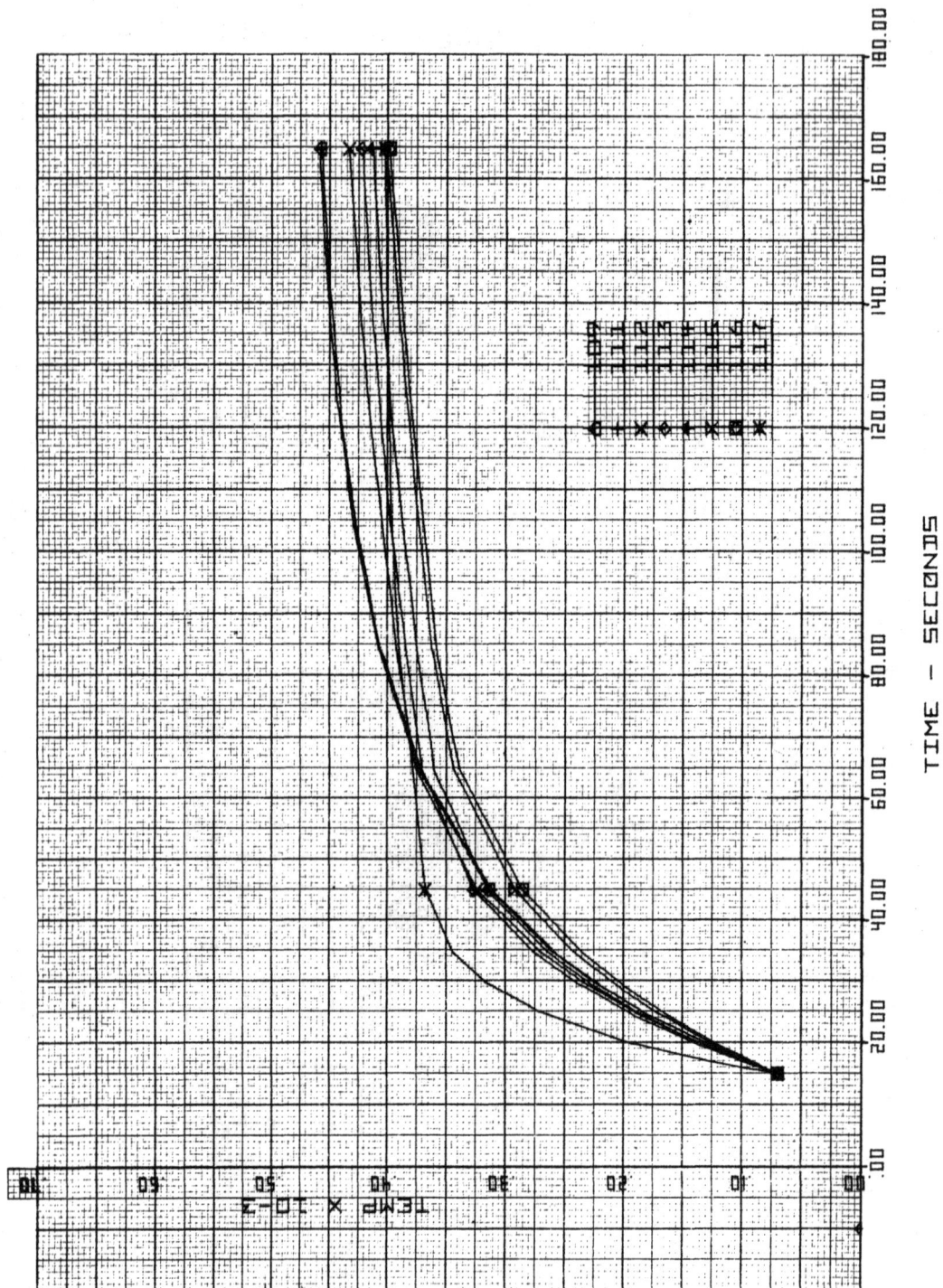

TIME - SECONDS

TEMP X 10⁻²

Figure A2. Forward Cooling Plate - Cold Start.

118

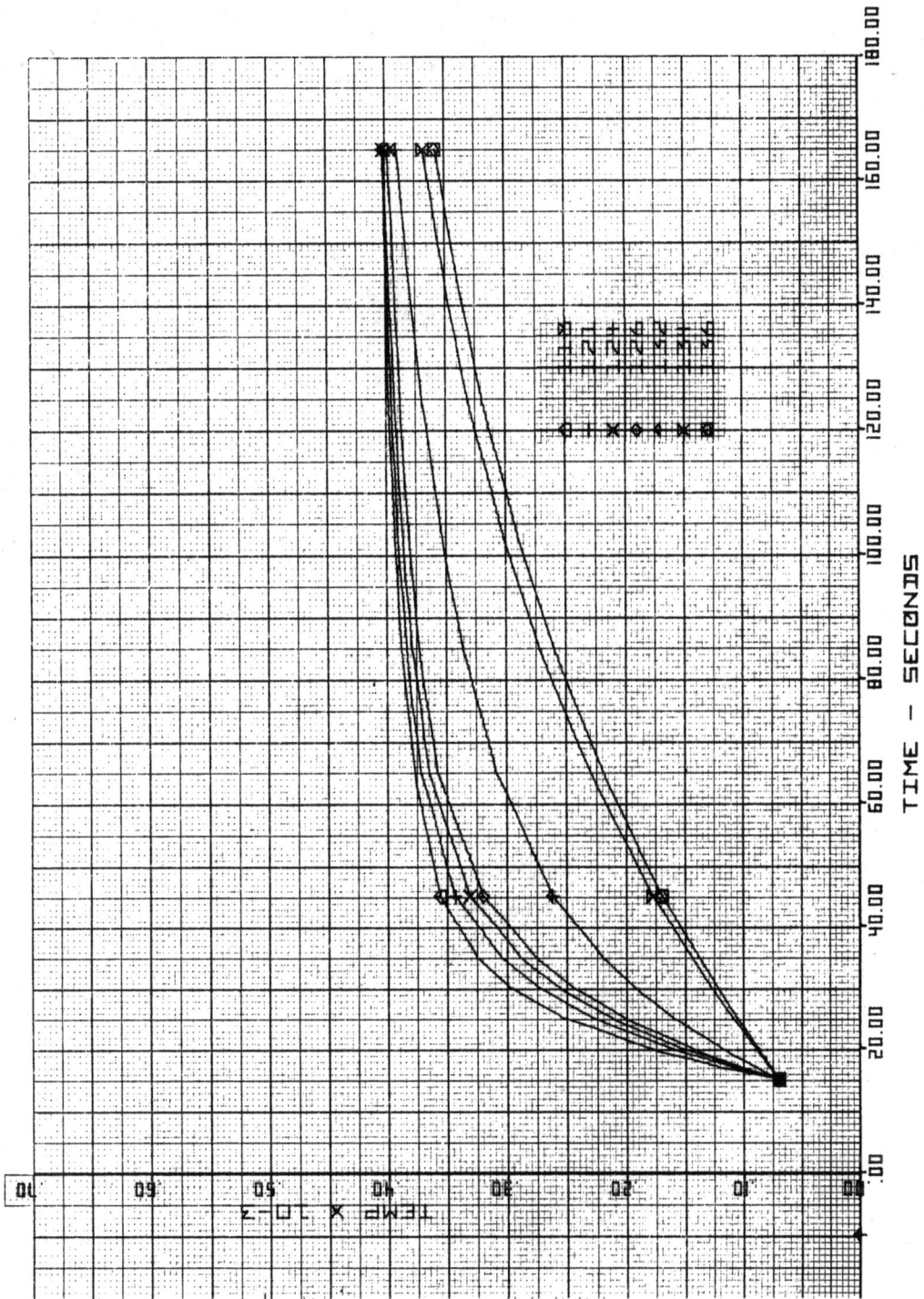

Figure A3. Forward Cooling Plate - Cold Start.

119

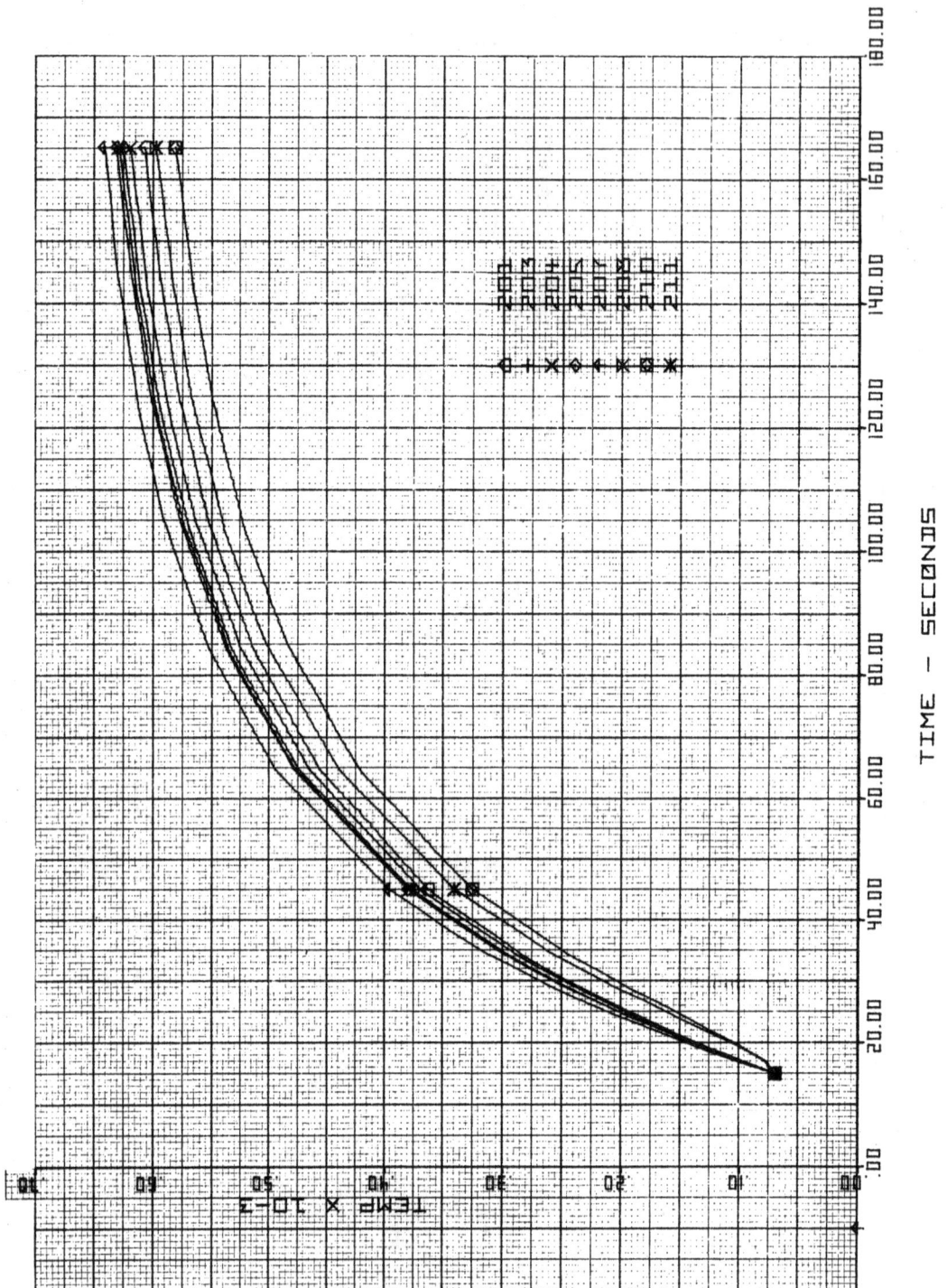

TIME – SECONDS

TEMP X 10-3

Figure A4. Aft Cooling Plate – Cold Start.

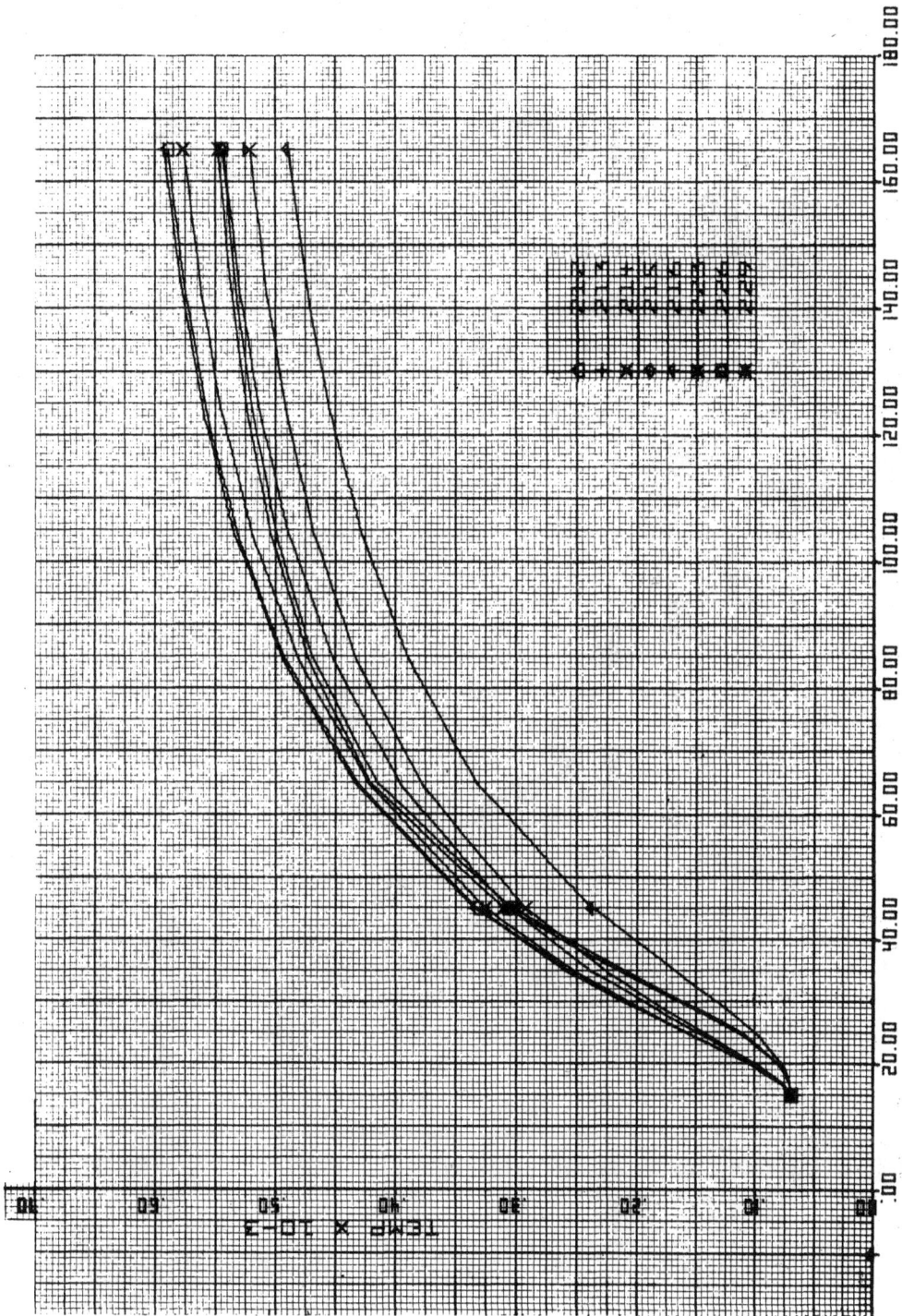

TIME - SECONDS

Figure A5. Aft Cooling Plate - Cold Start.

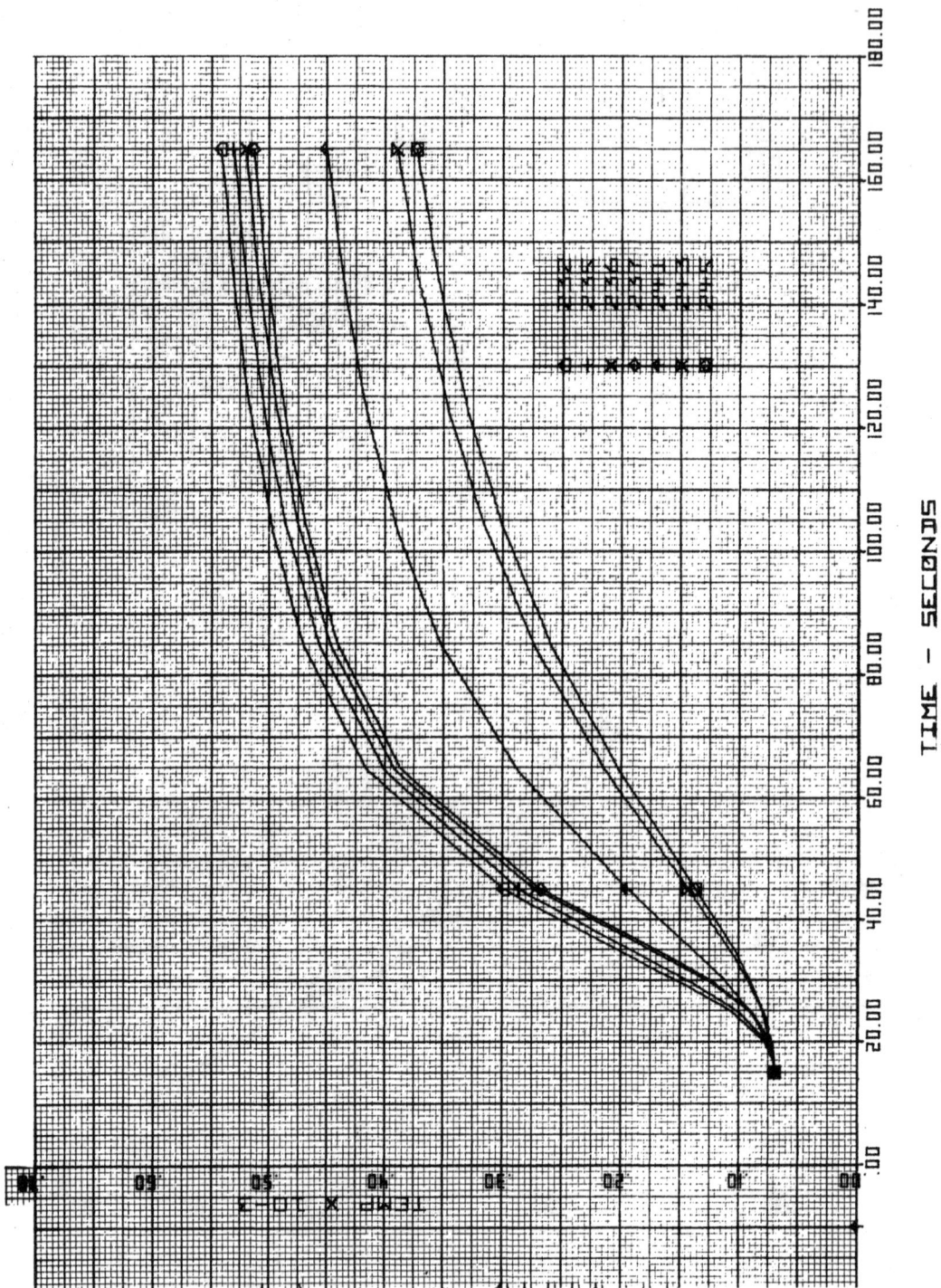

TIME - SECONDS

Figure A6. Aft Cooling Plate - Cold Start.

122

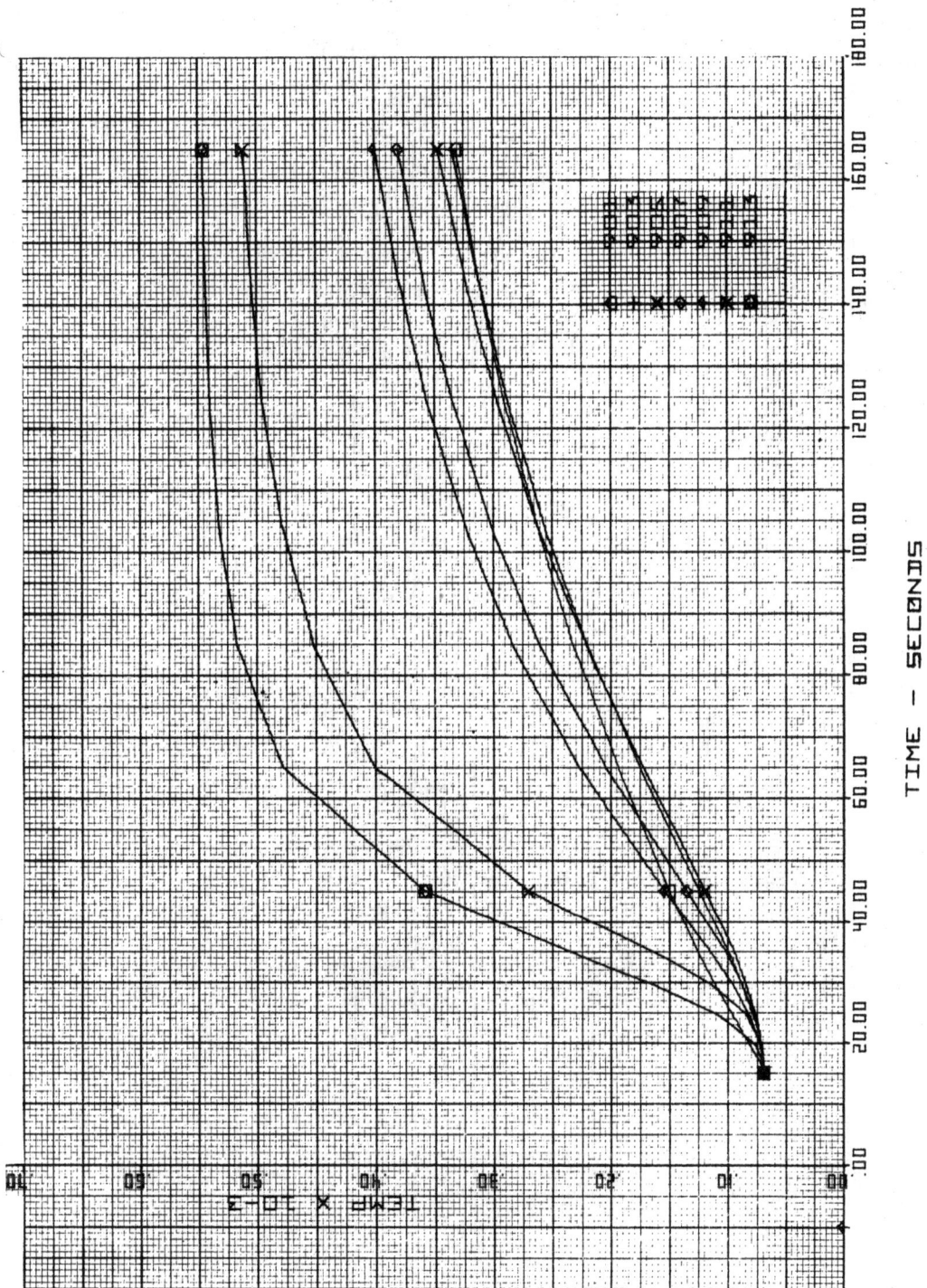

Figure A7. Disk - Cold Start.

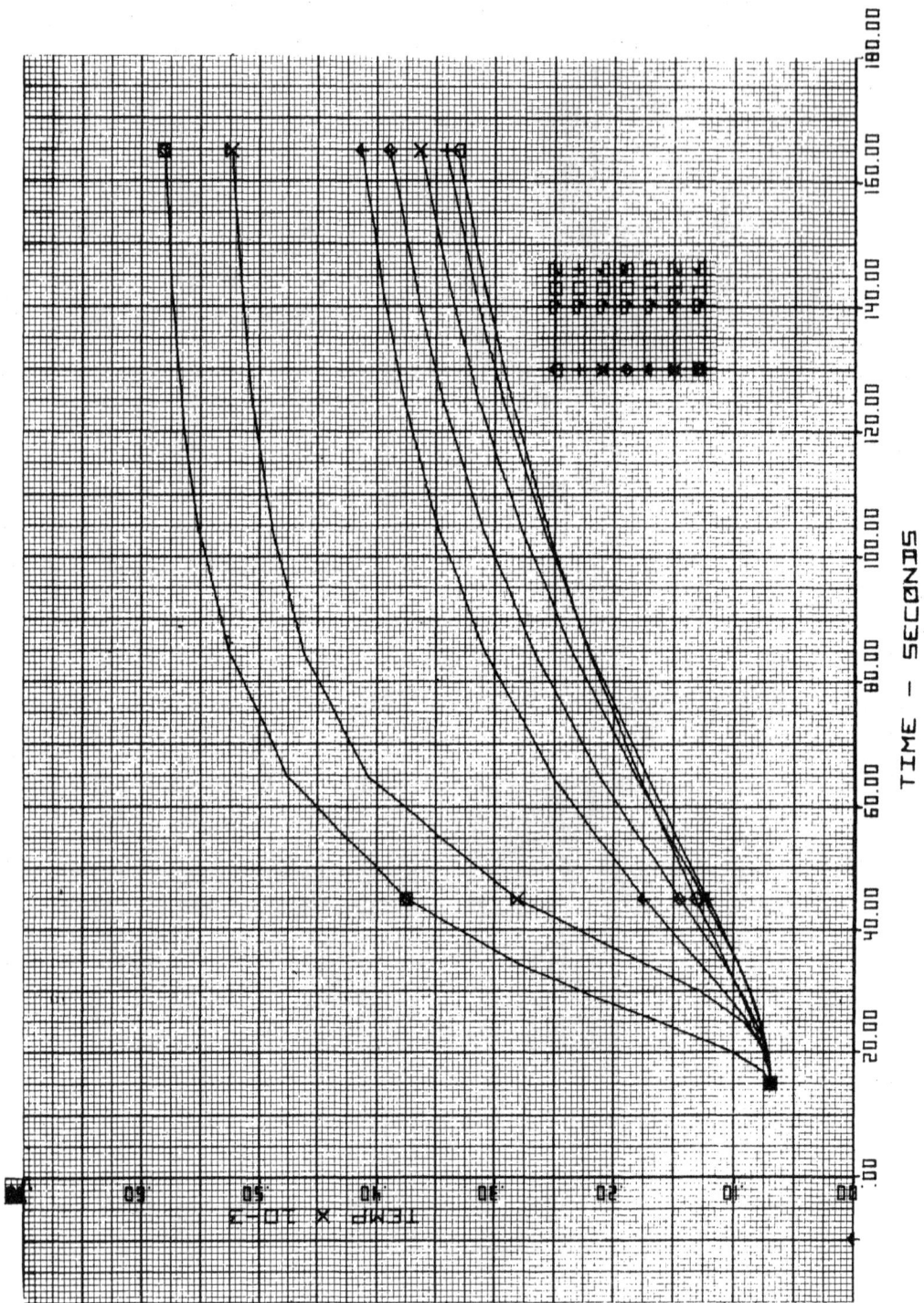

Figure A8. Disk - Cold Start.

124

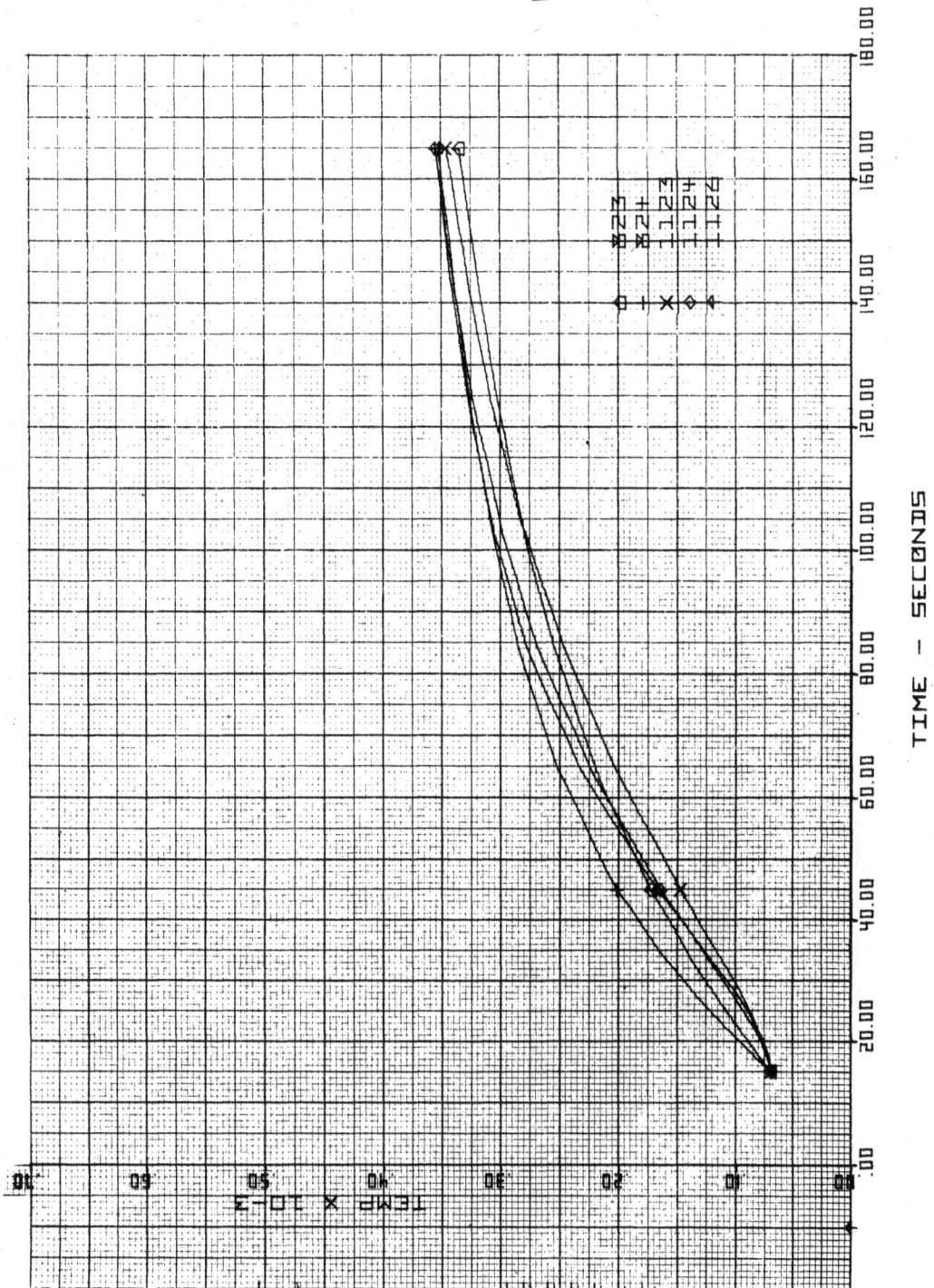

TIME – SECONDS

Figure A9. Miscellaneous Nodes – Cold Start.

125

A10. Forward Cooling Plate - Burst.

FWD COOLING PLATE - BURST

TEMPERATURE - DEG. F *10¹

TIME - SECONDS

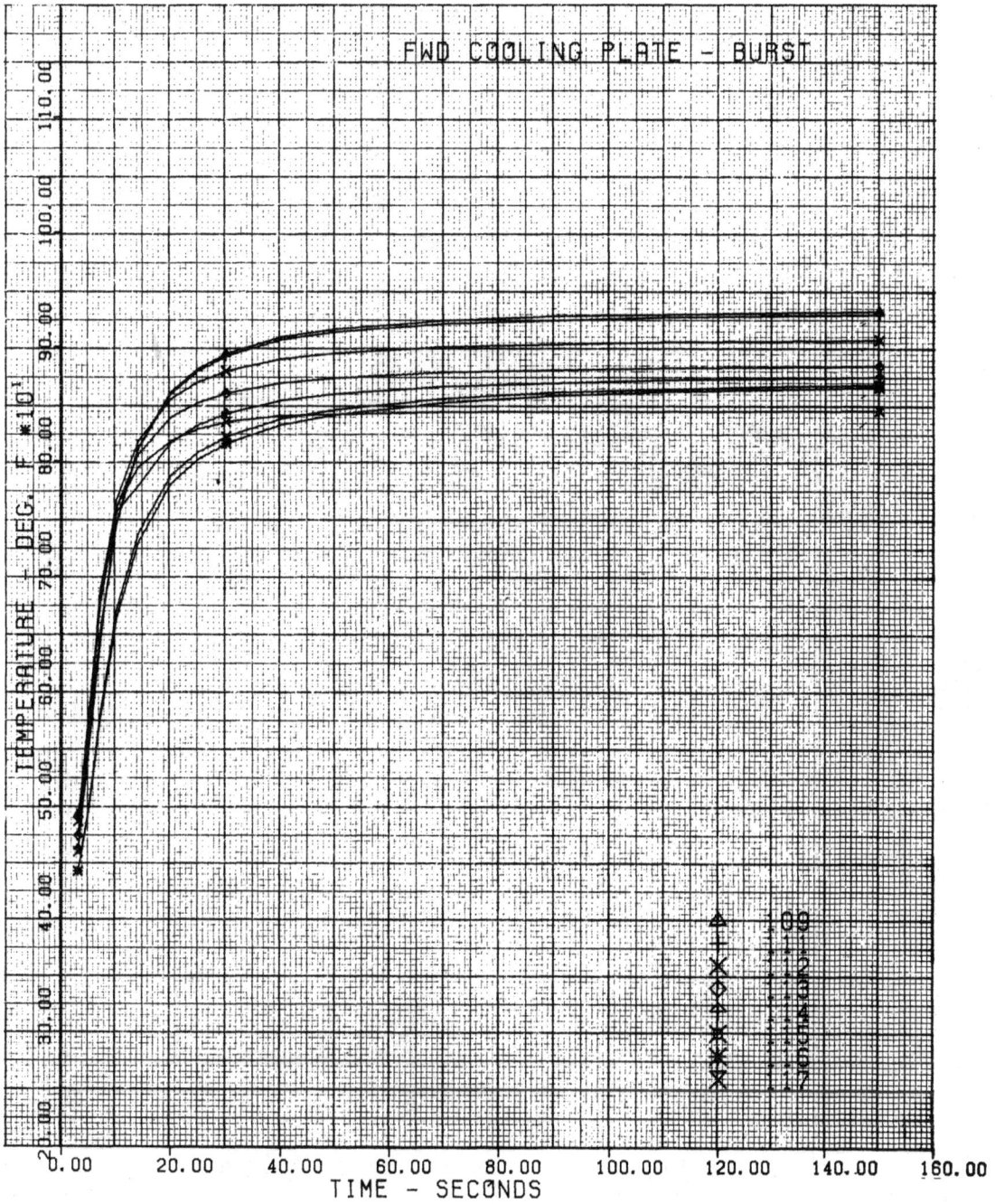

A11. Forward Cooling Plate - Burst.

127

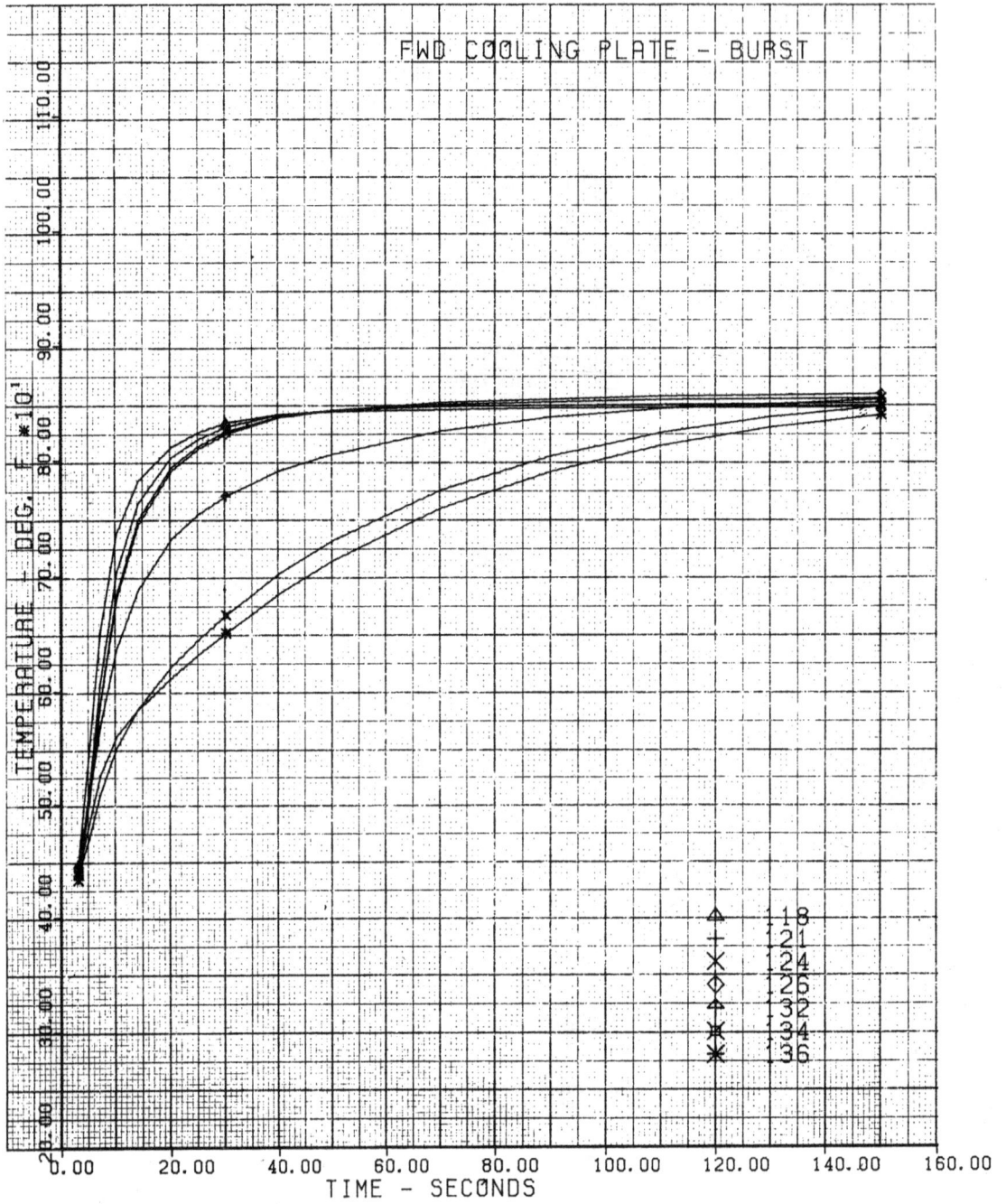

FWD COOLING PLATE - BURST

A12. Forward Cooling Plate - Burst.

128

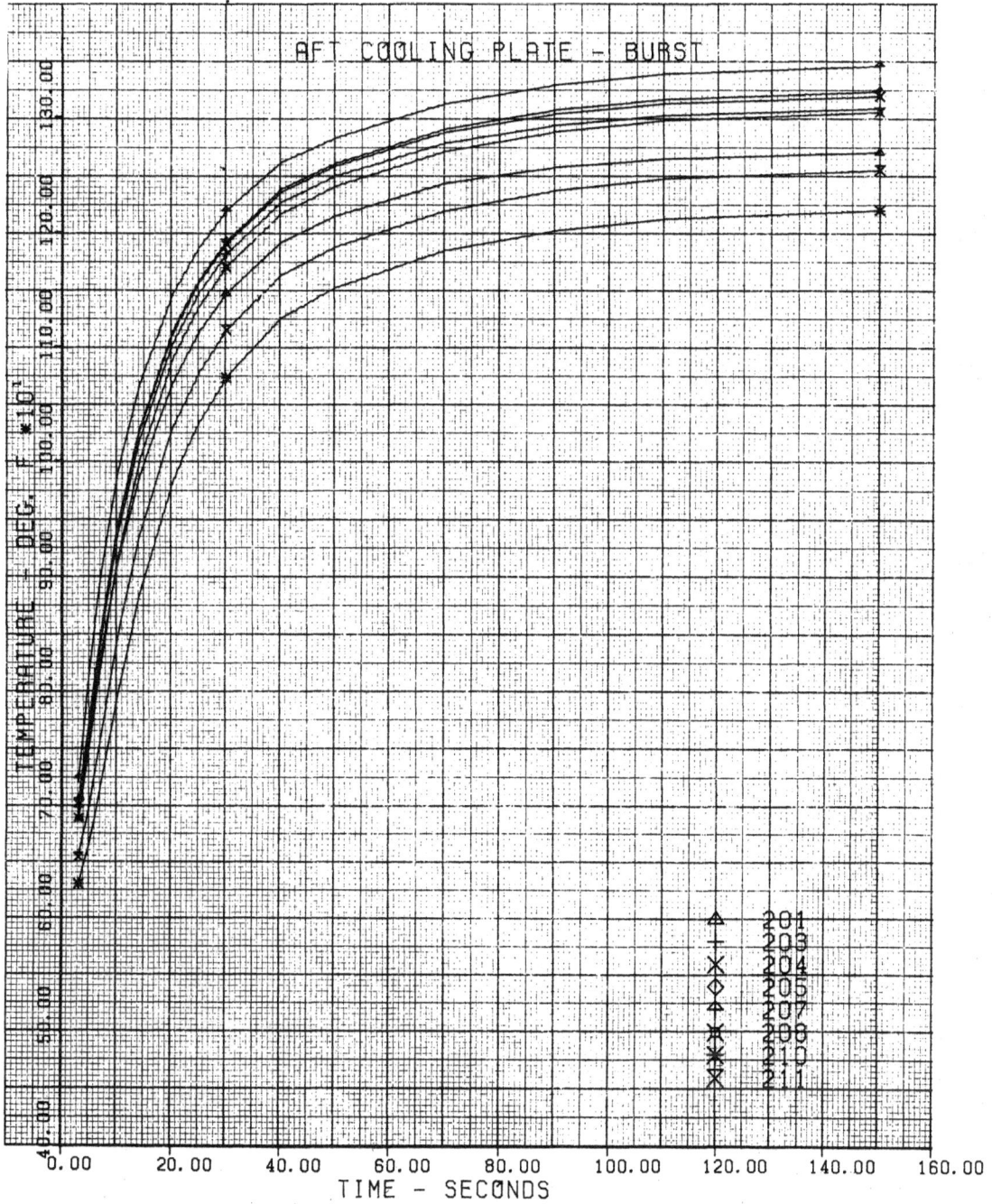

A13. Aft Cooling Plate - Burst.

129

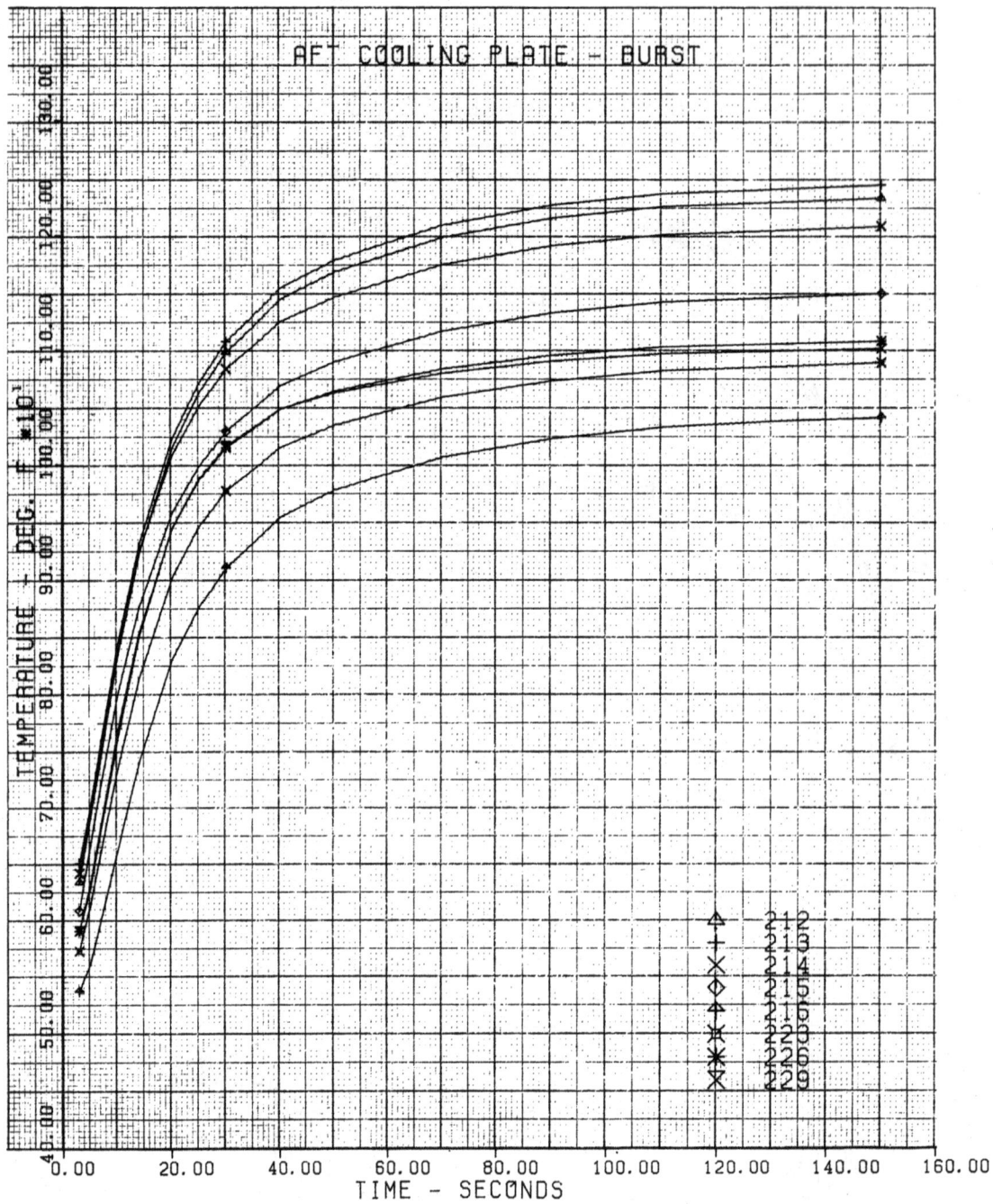

AFT COOLING PLATE - BURST

A14.   Aft Cooling Plate - Burst.

130

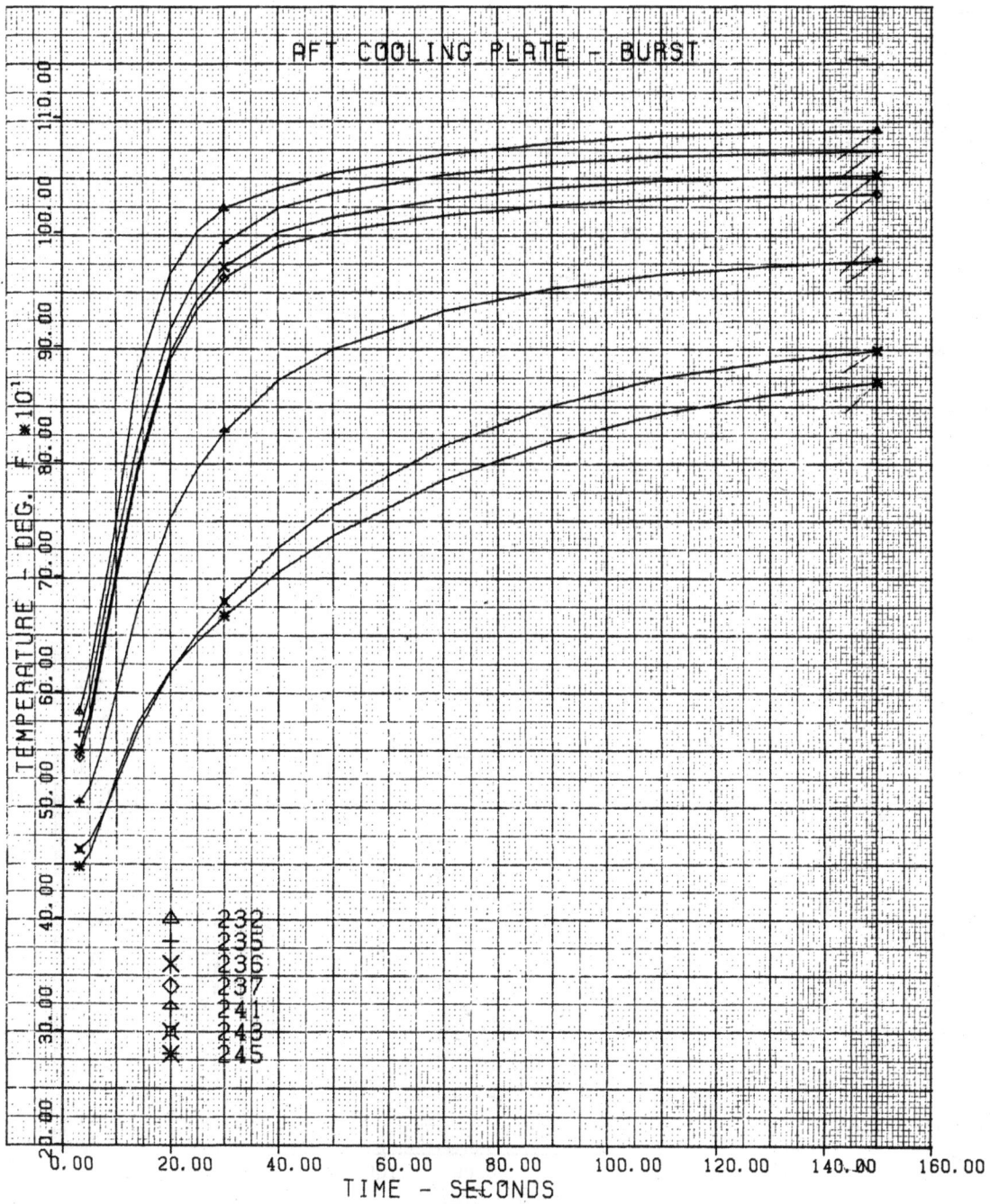

AFT COOLING PLATE - BURST

TEMPERATURE - DEG. F *10¹

TIME - SECONDS

Legend:
△ 232
+ 235
× 236
◇ 237
⊿ 241
✳ 243
✻ 245

A15. Aft Cooling Plate - Burst.

131

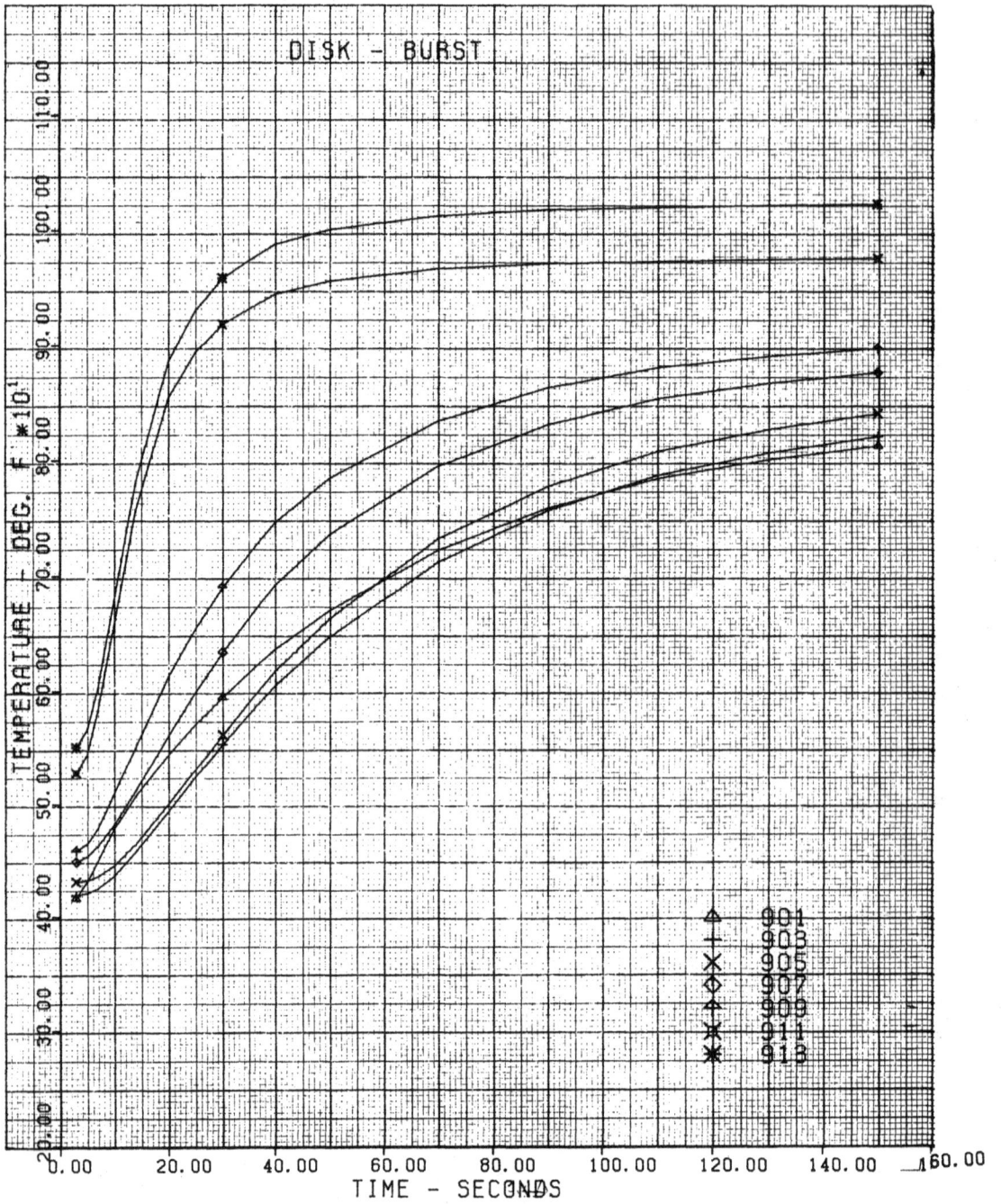

A16. Disk - Burst.

132

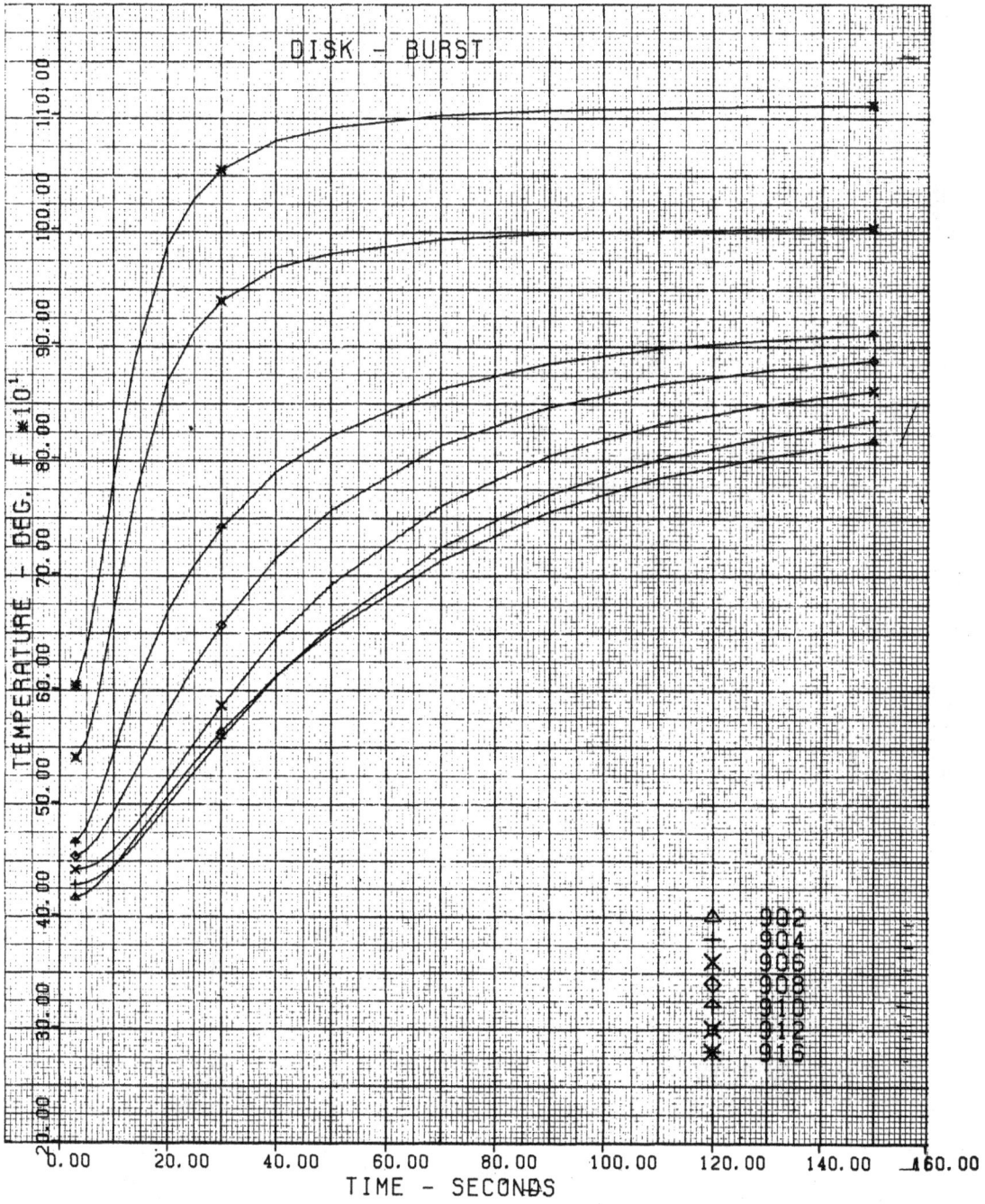

A17.   Disk — Burst.

133

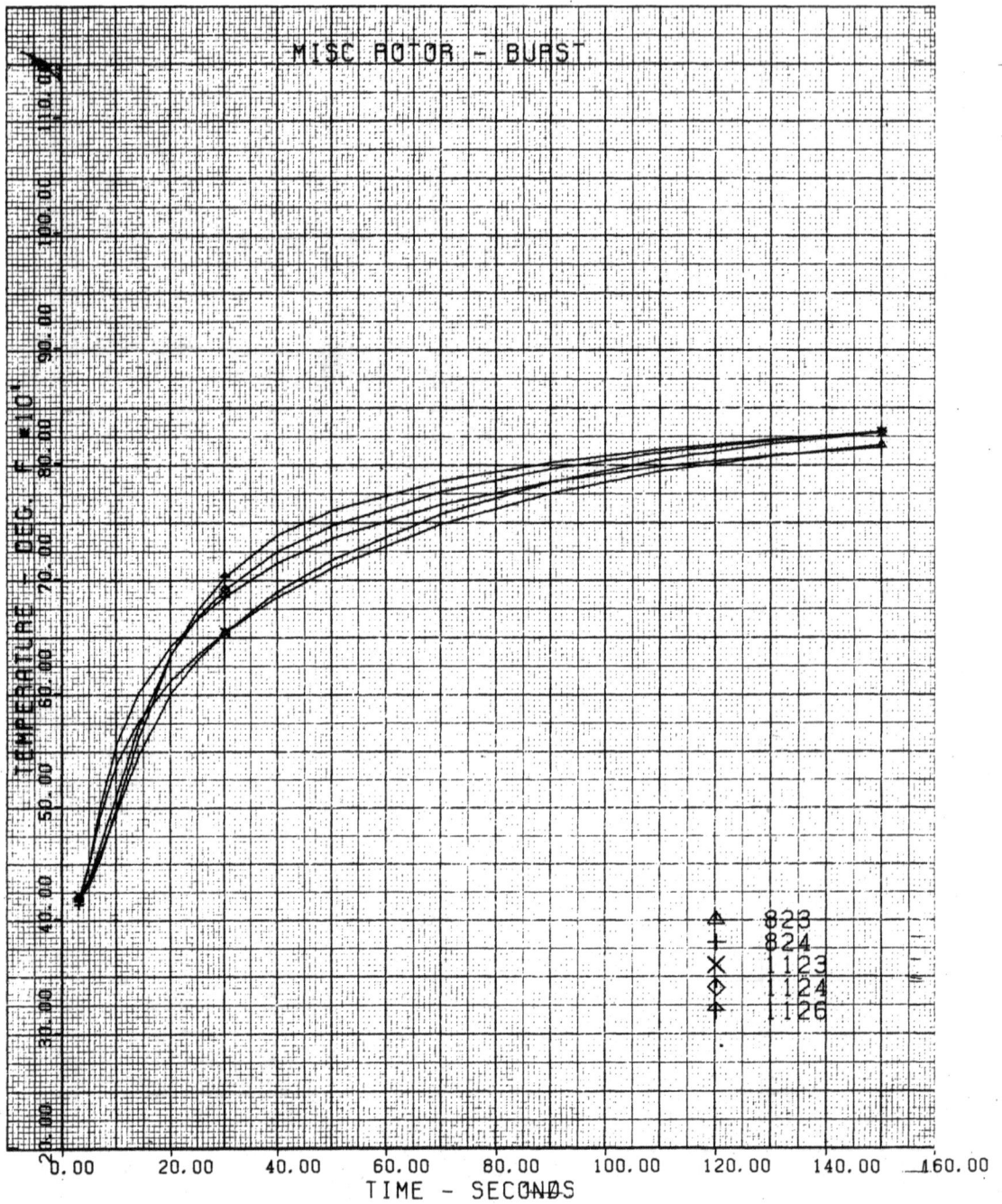

A18. Micsellaneous Nodes - Burst.

134

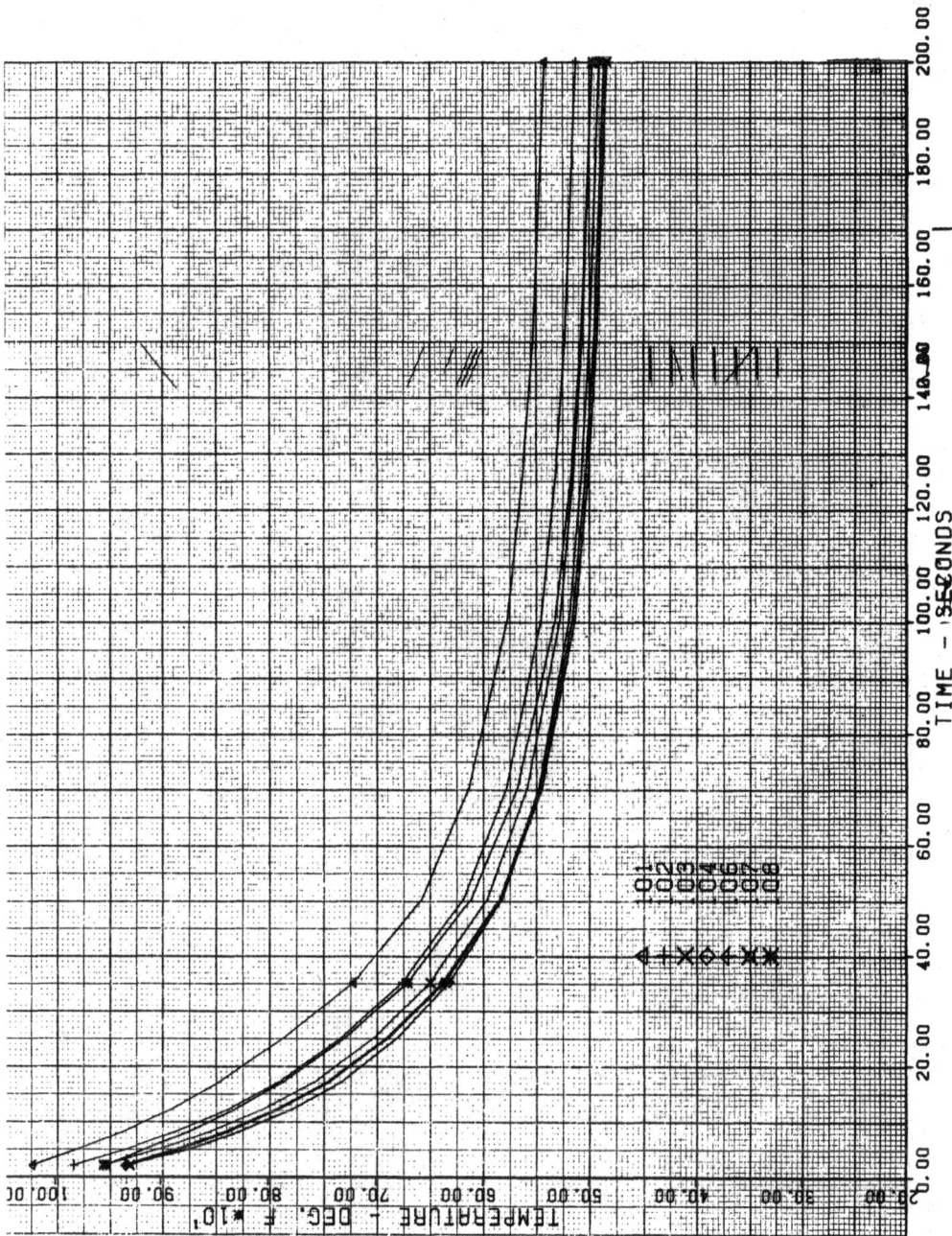

A19. Forward Cooling Plate – Chop.

135

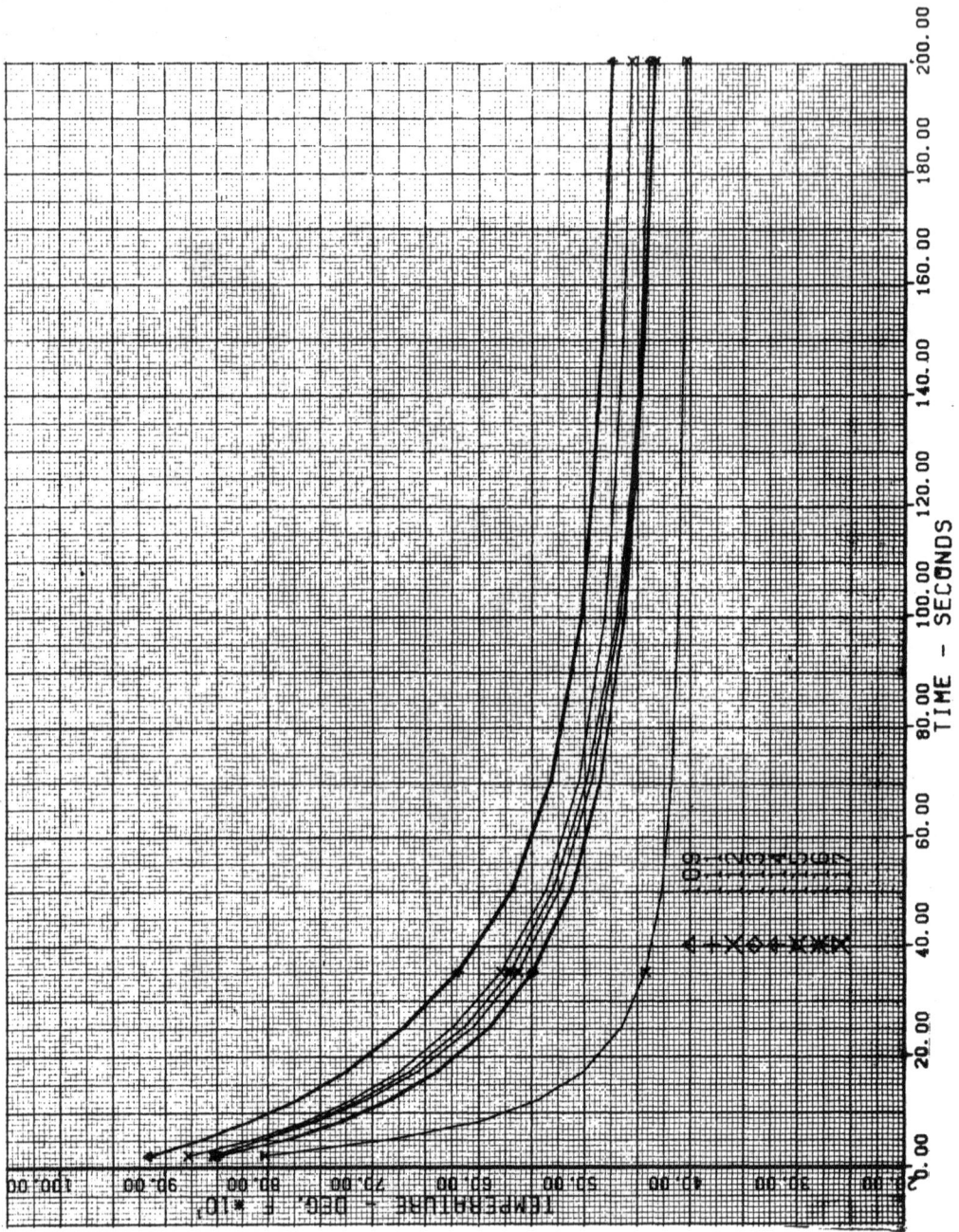

A20. Forward Cooling Plate — Chop.

136

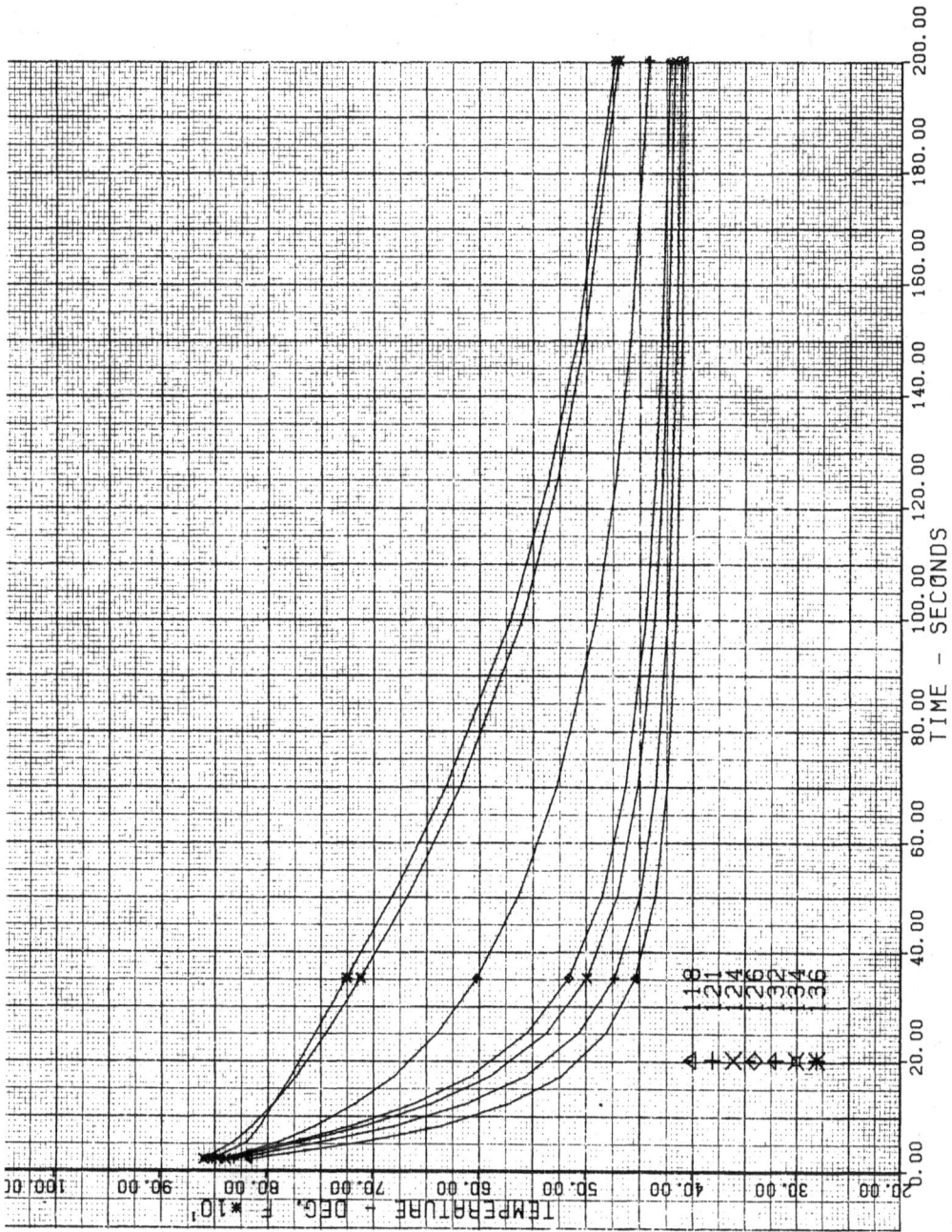

A21. Forward Cooling Plate – Chop.

137

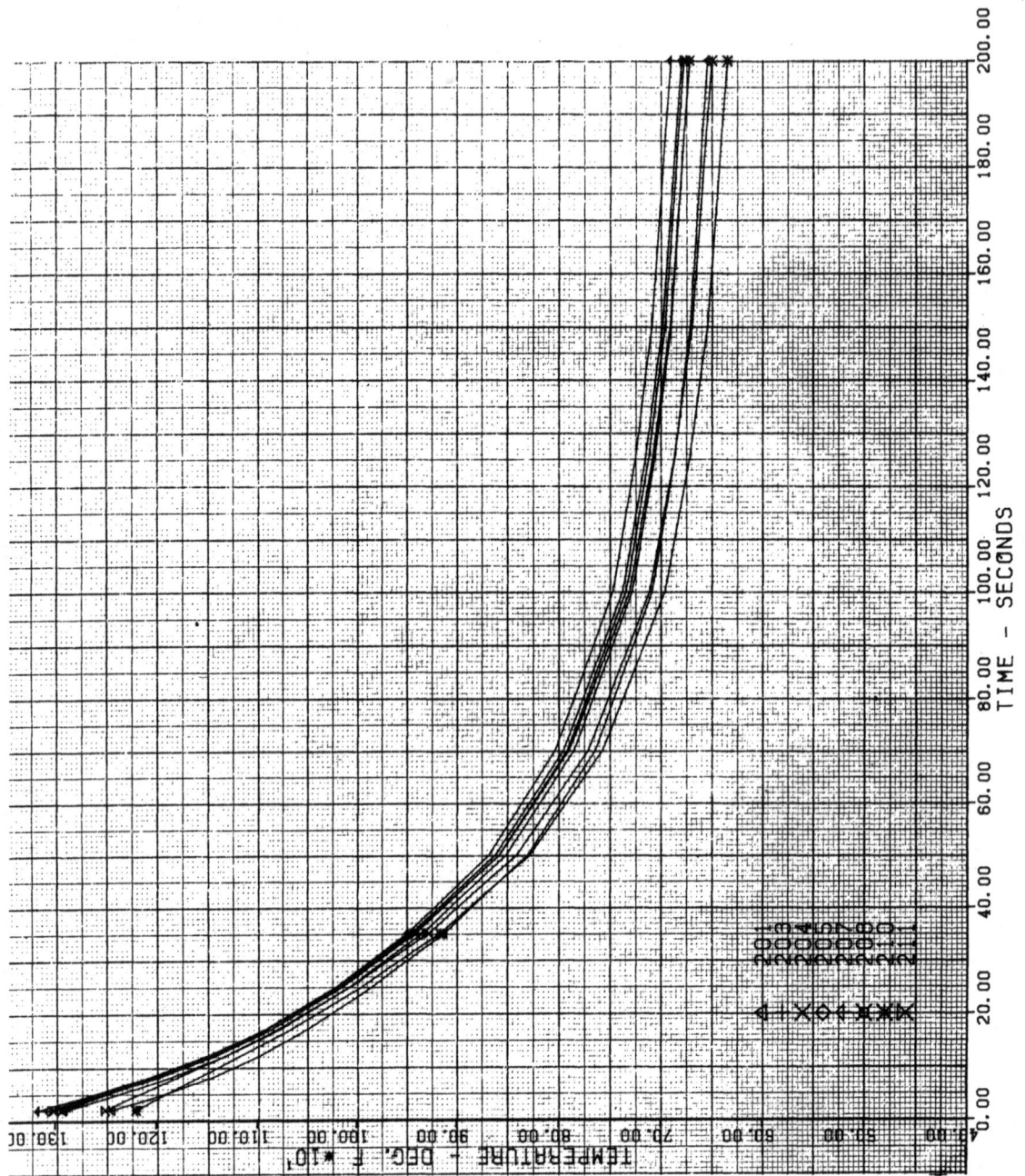

A22. Aft Cooling Plate - Chop.

138

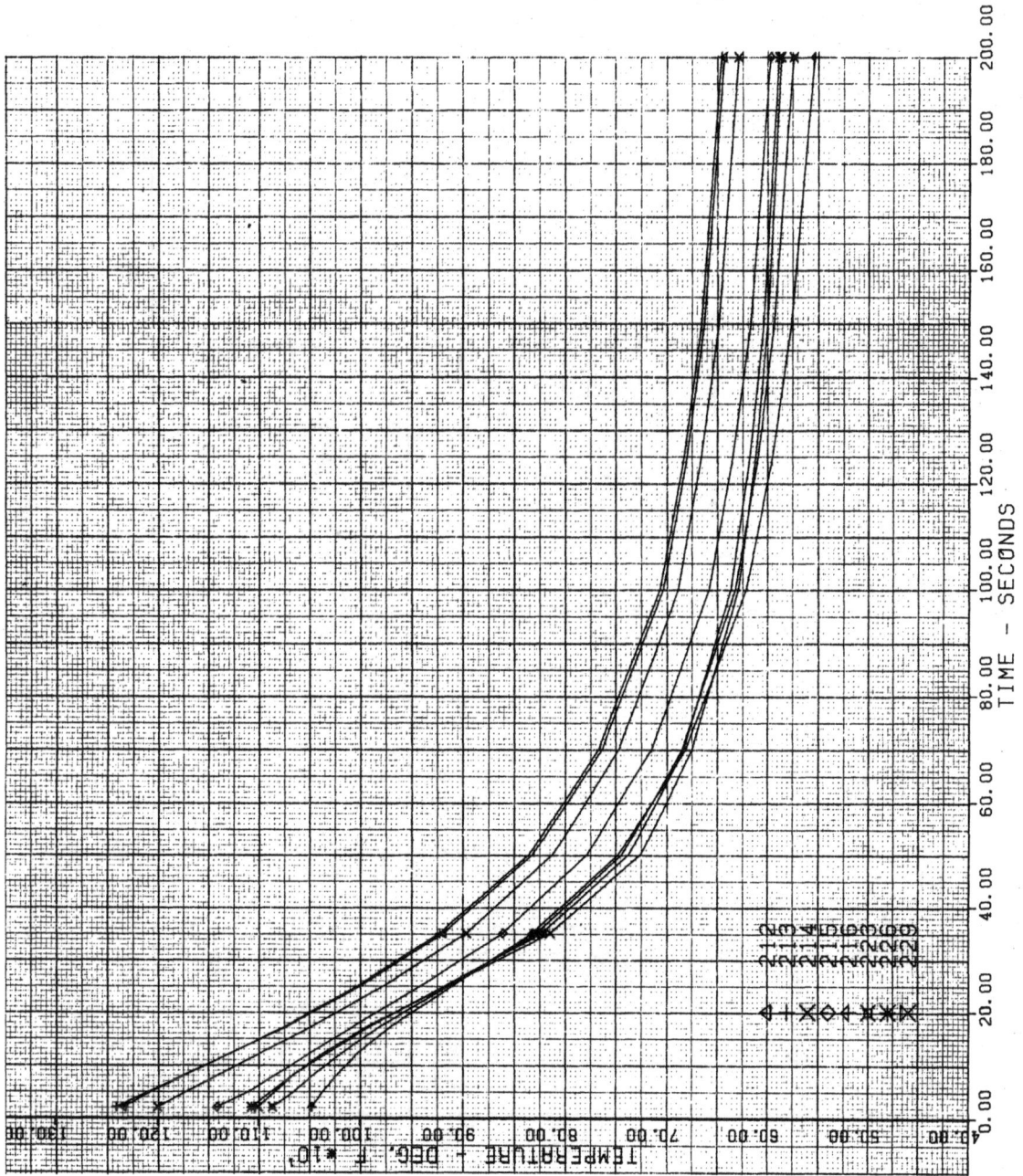

A23. Aft Cooling Plate - Chop.

139

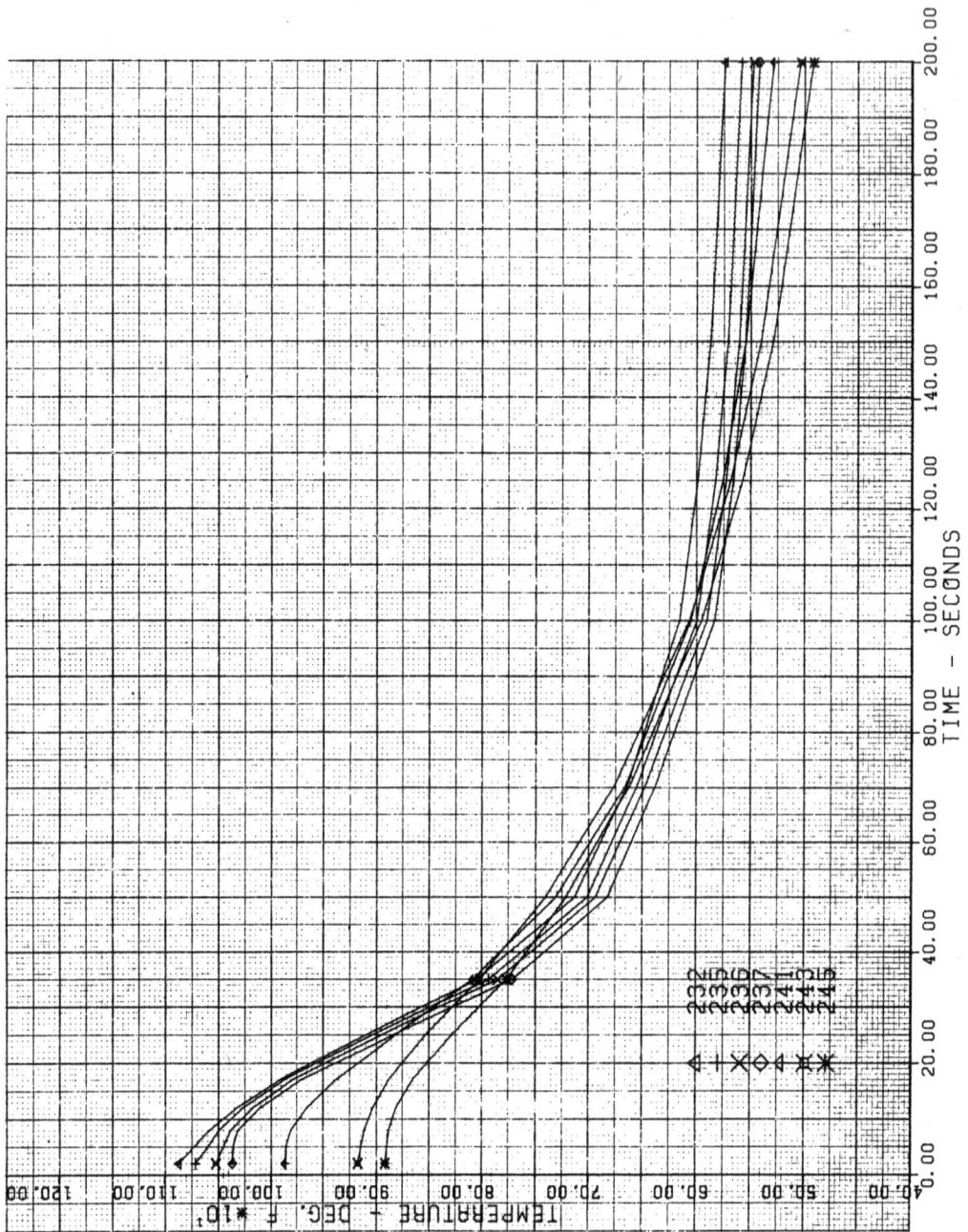

A24. Aft Cooling Plate – Chop.

A25. Disk - Chop.

141

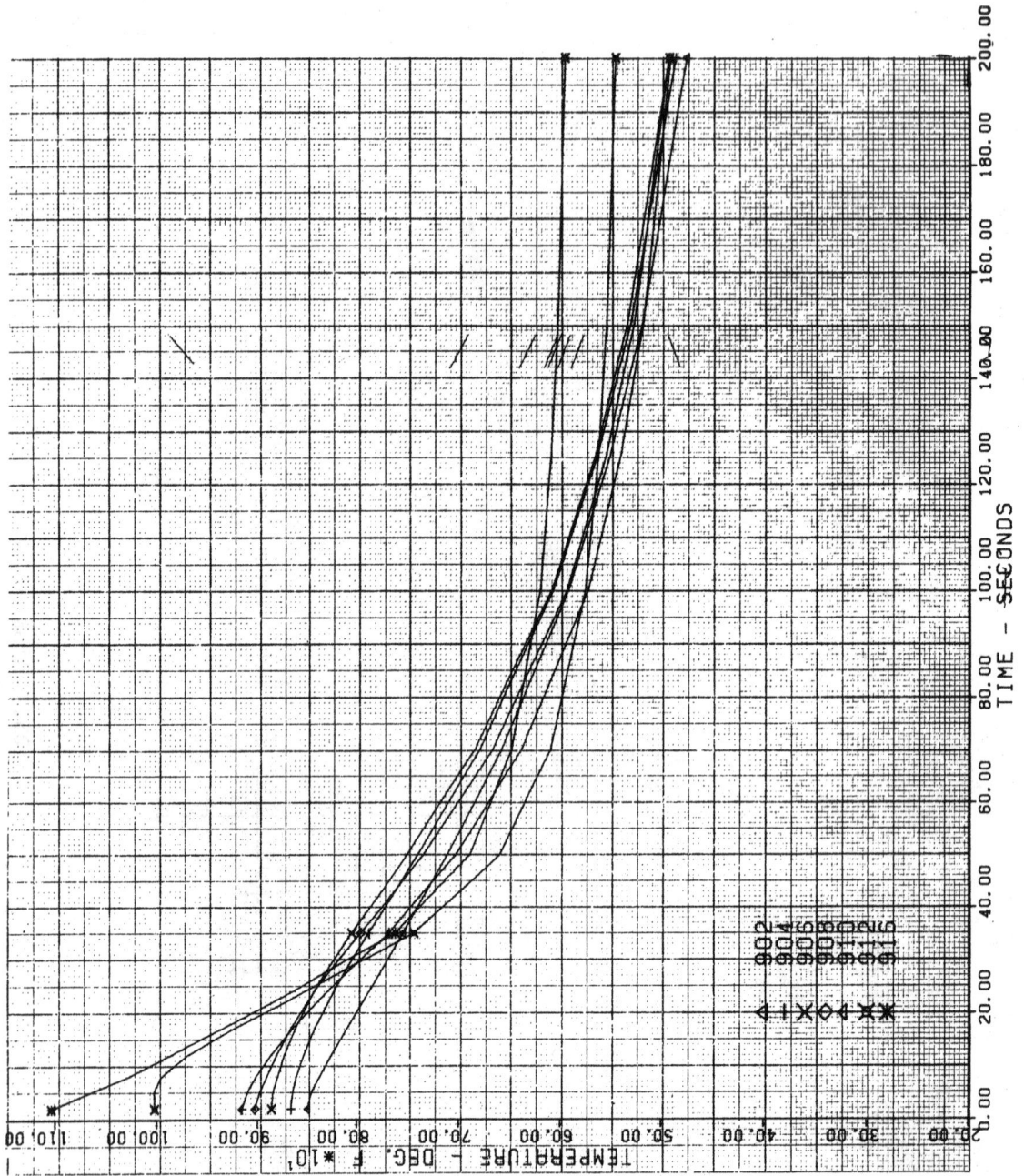

A26. Disk - Chop.

142

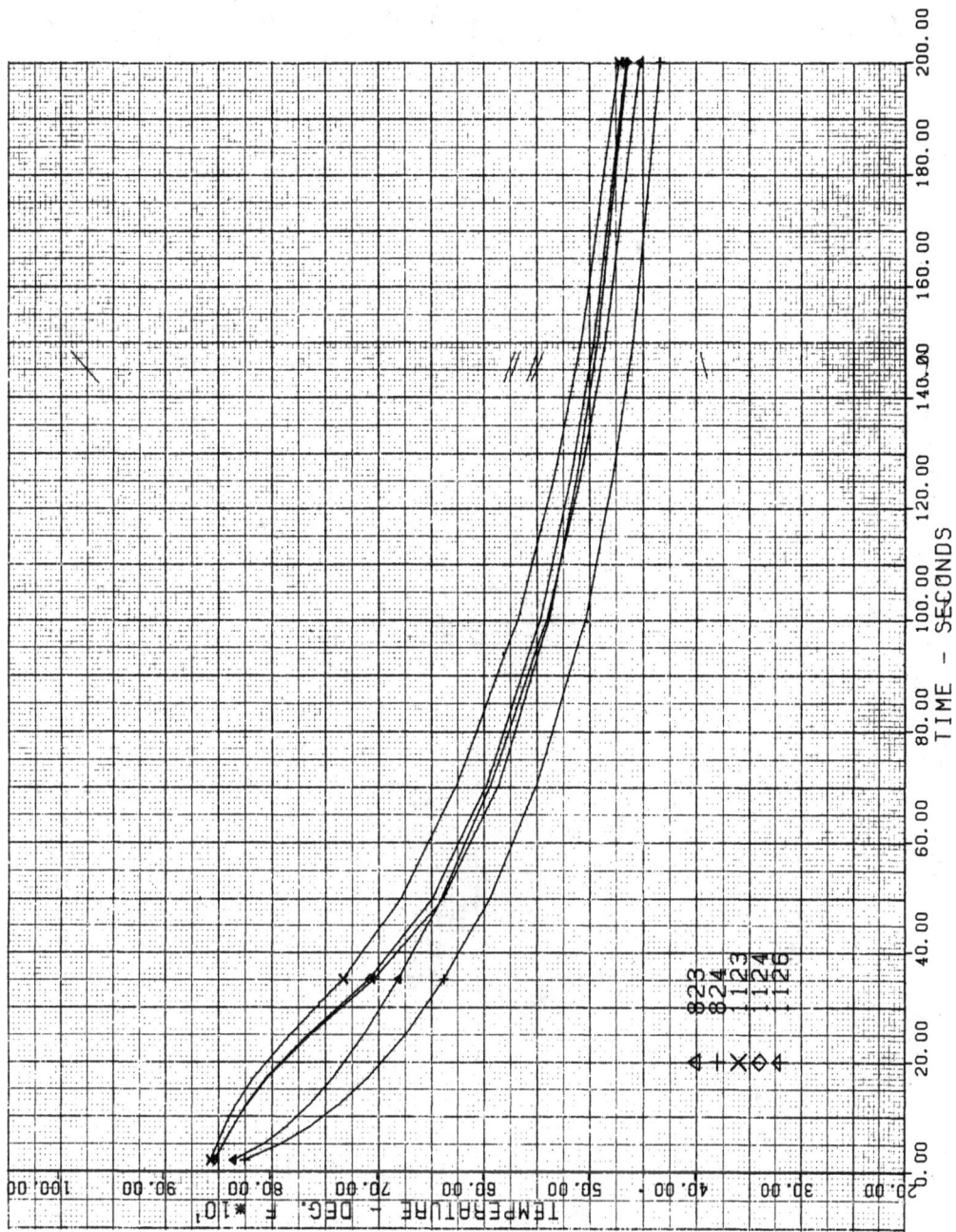

A27. Miscellaneous Nodes – Chop.

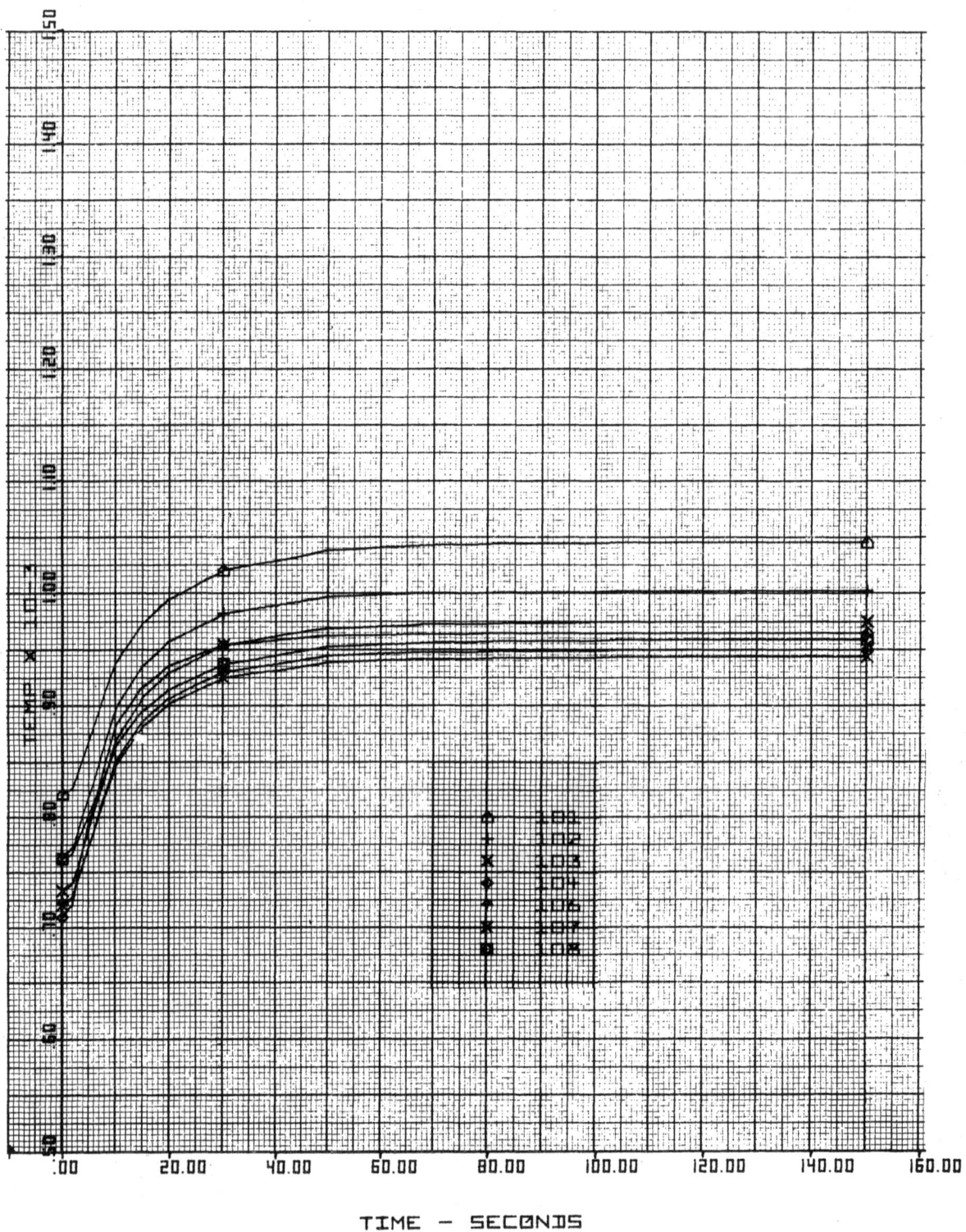

Figure A28. Forward Cooling Plate - Reburst.

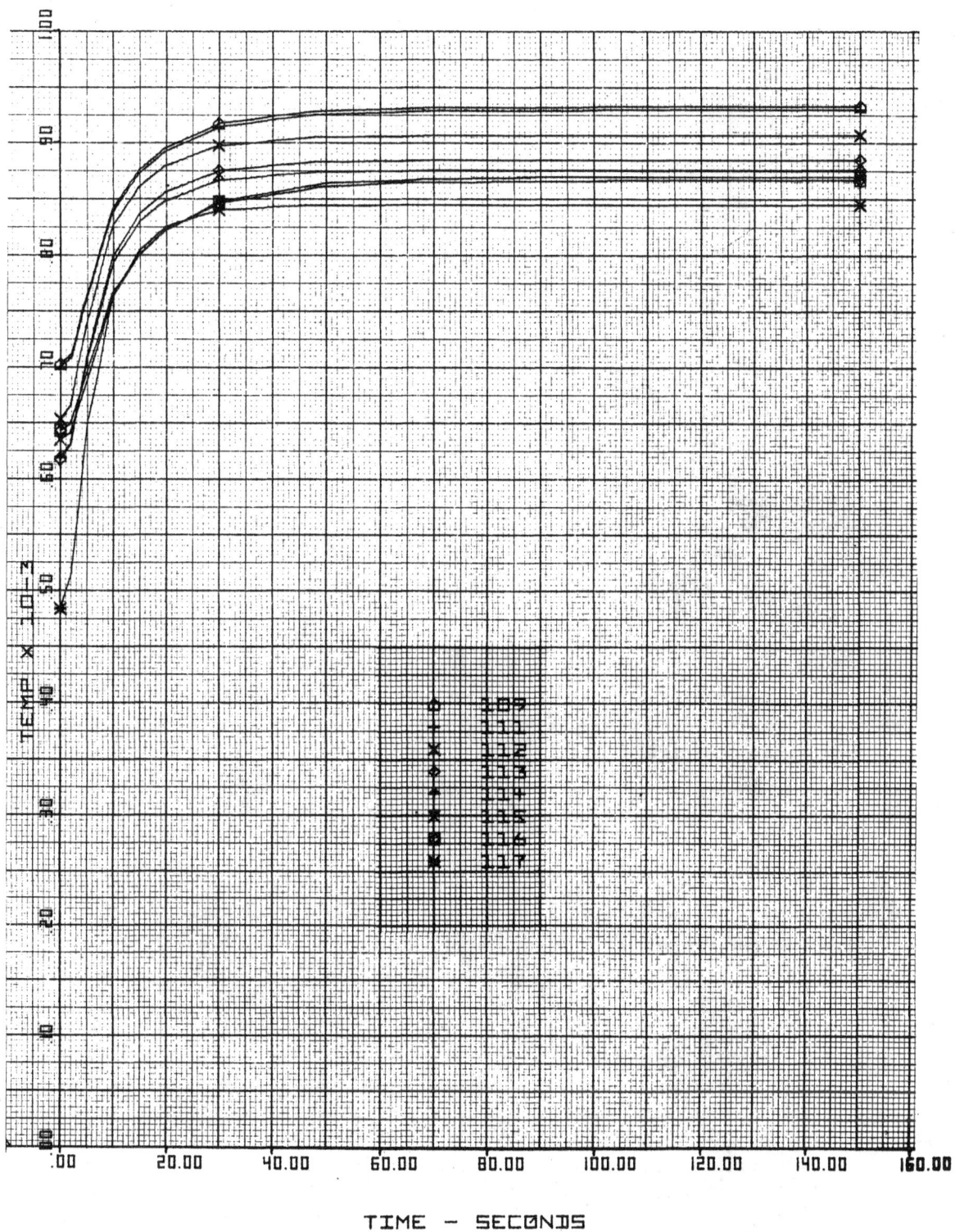

Figure A29. Forward Cooling Plate - Reburst.

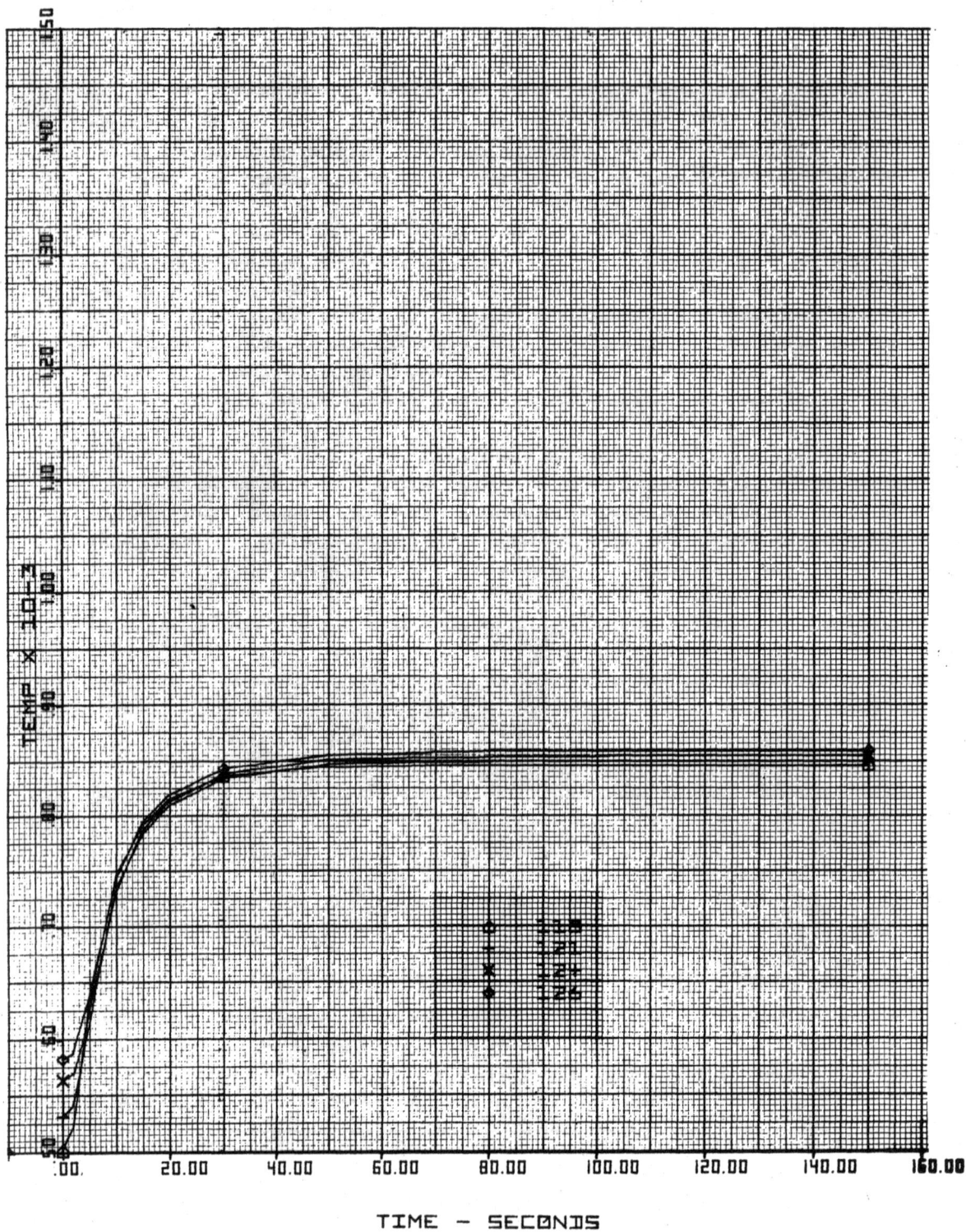

Figure A30.   Forward Cooling Plate - Reburst.

146

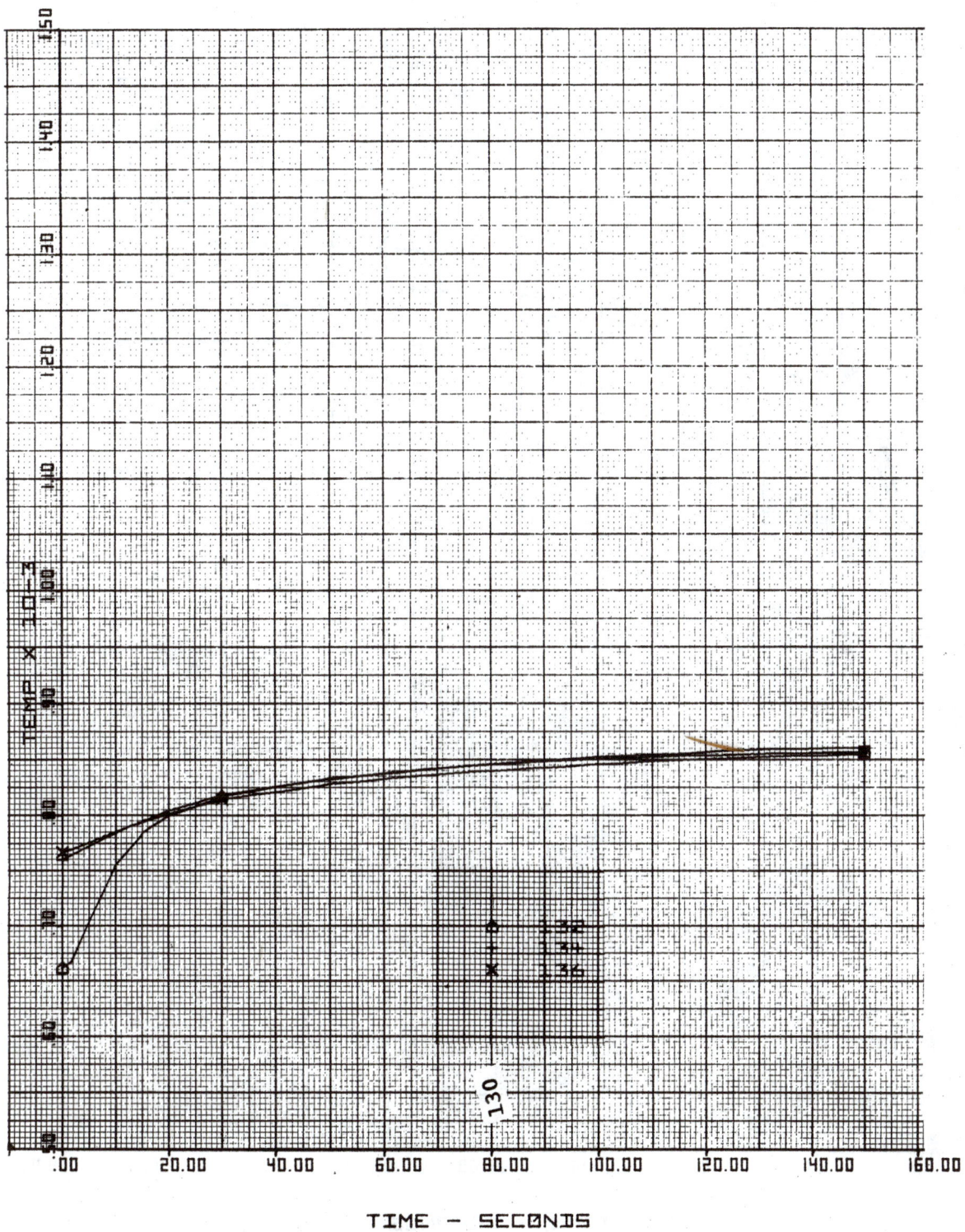

Figure A31. Forward Cooling Plate - Reburst.

147

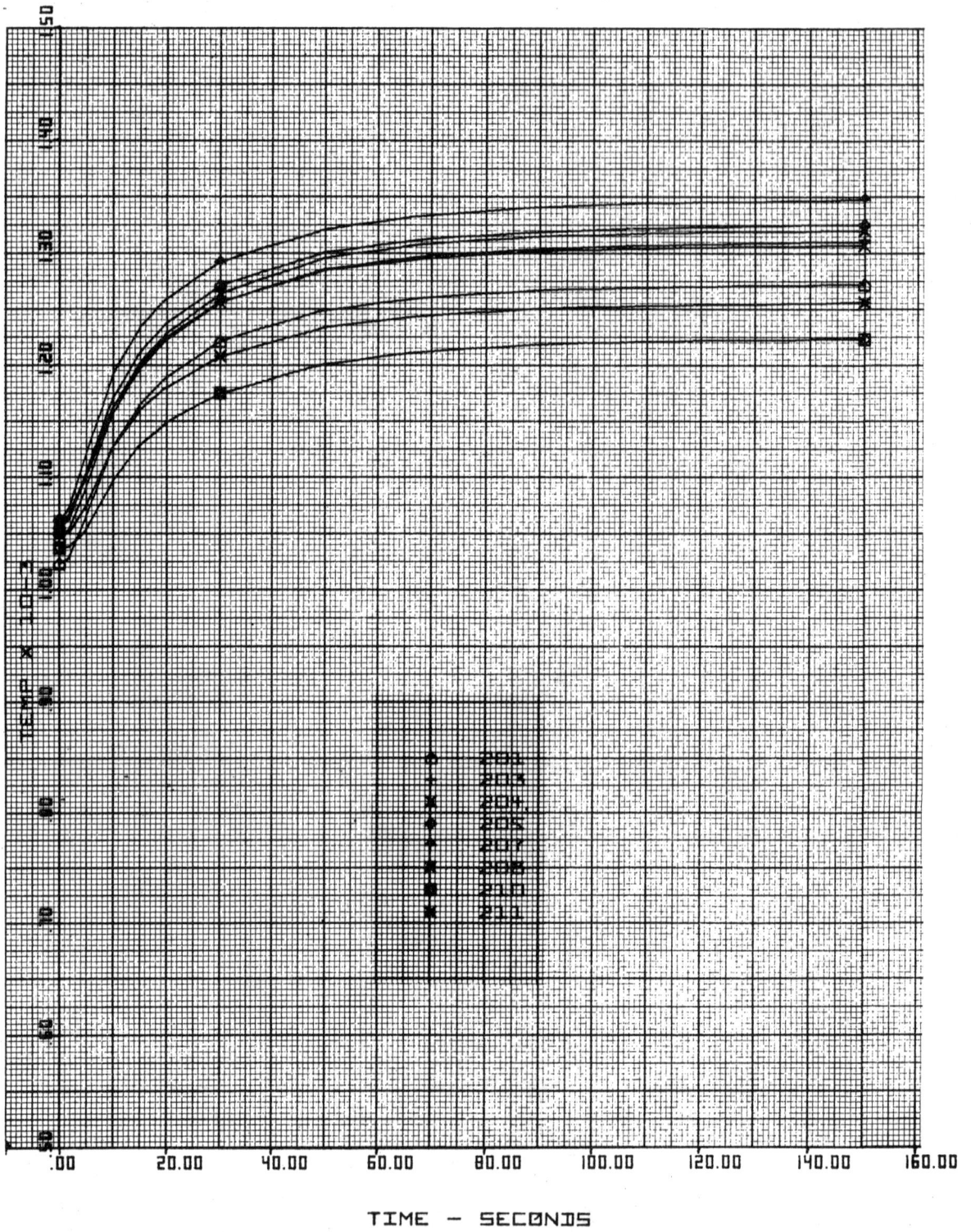

Figure A32.   Aft Cooling Plate - Reburst.

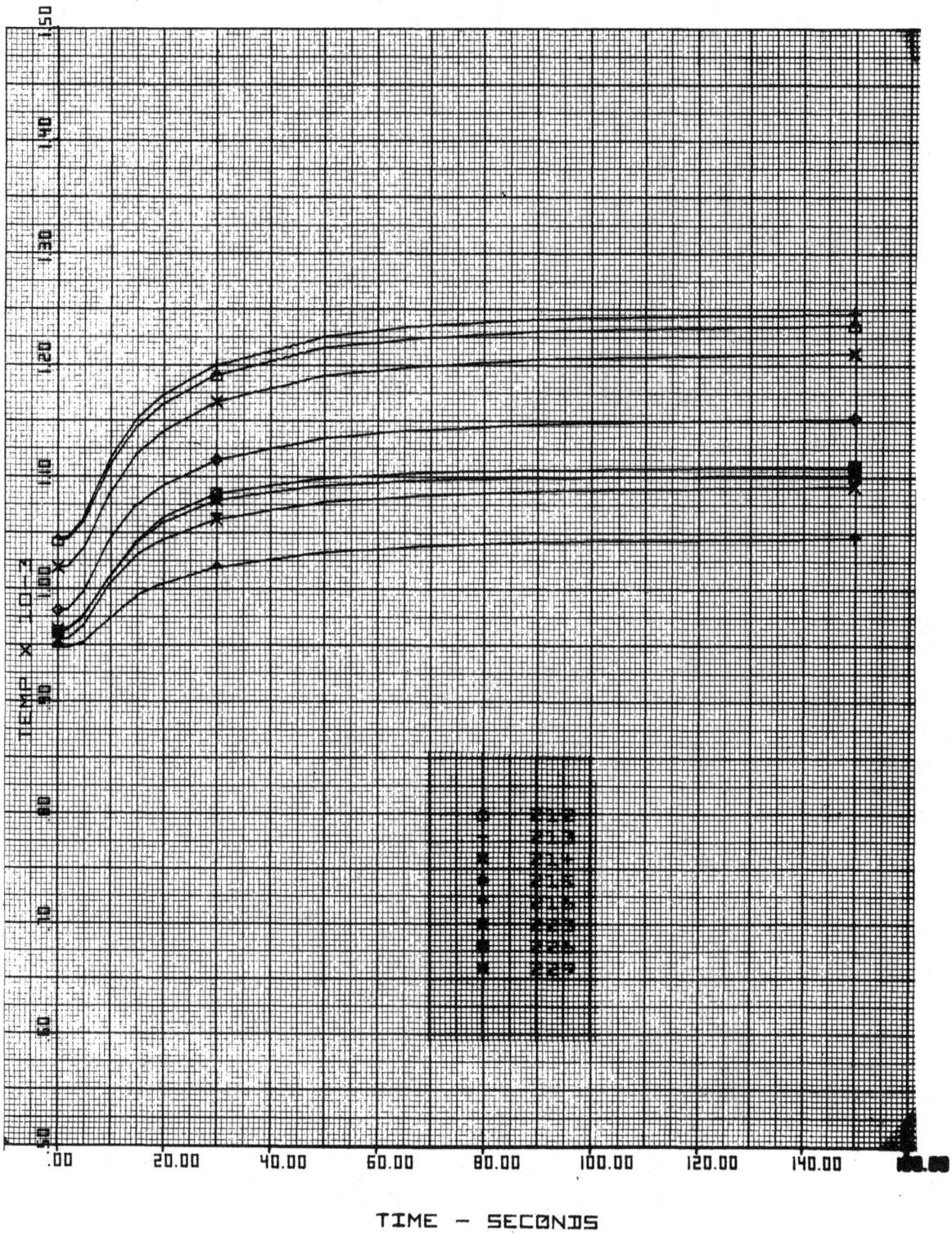

Figure A33. Aft Cooling Plate - Reburst.

149

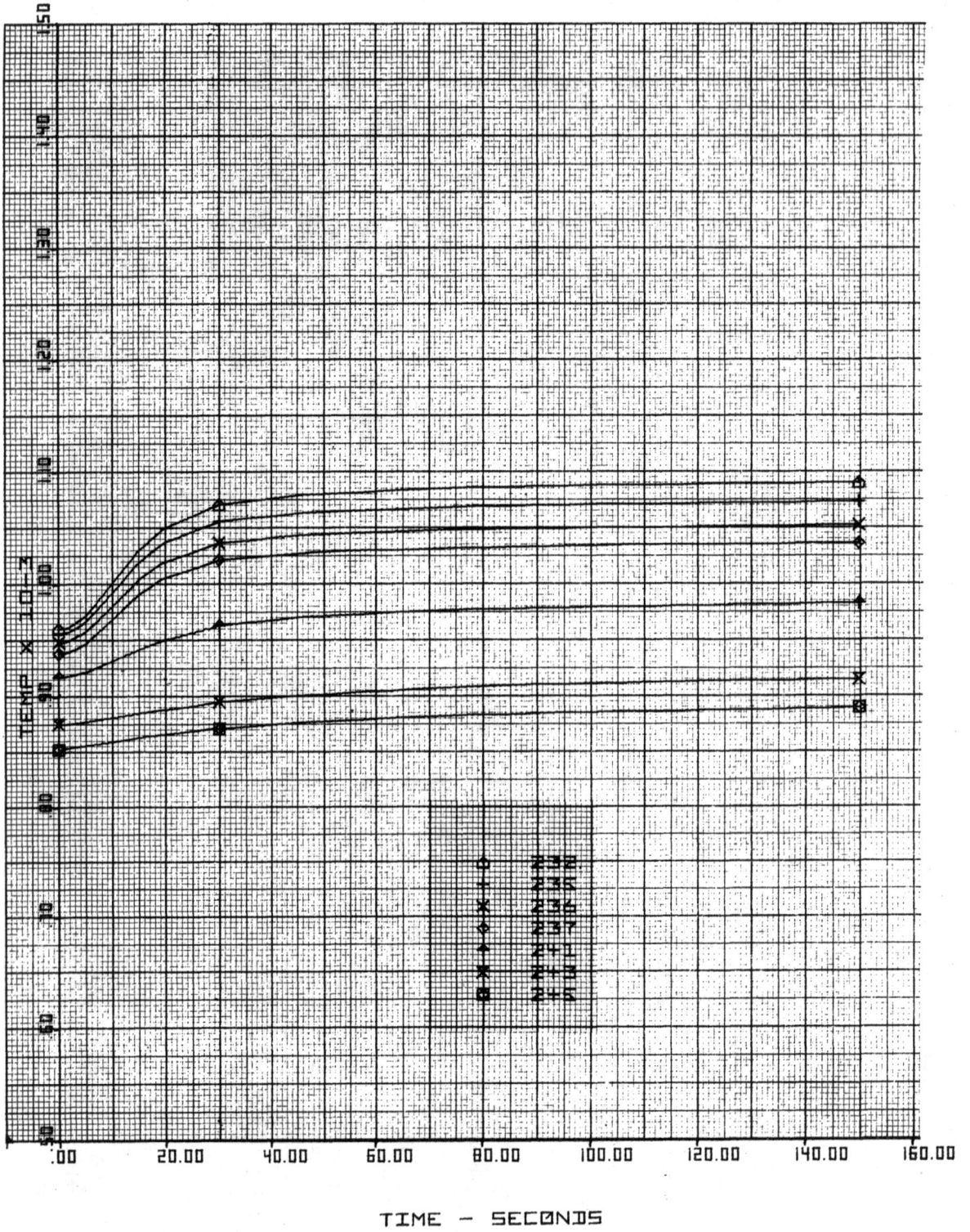

Figure A34. Aft Cooling Plate - Reburst.

150

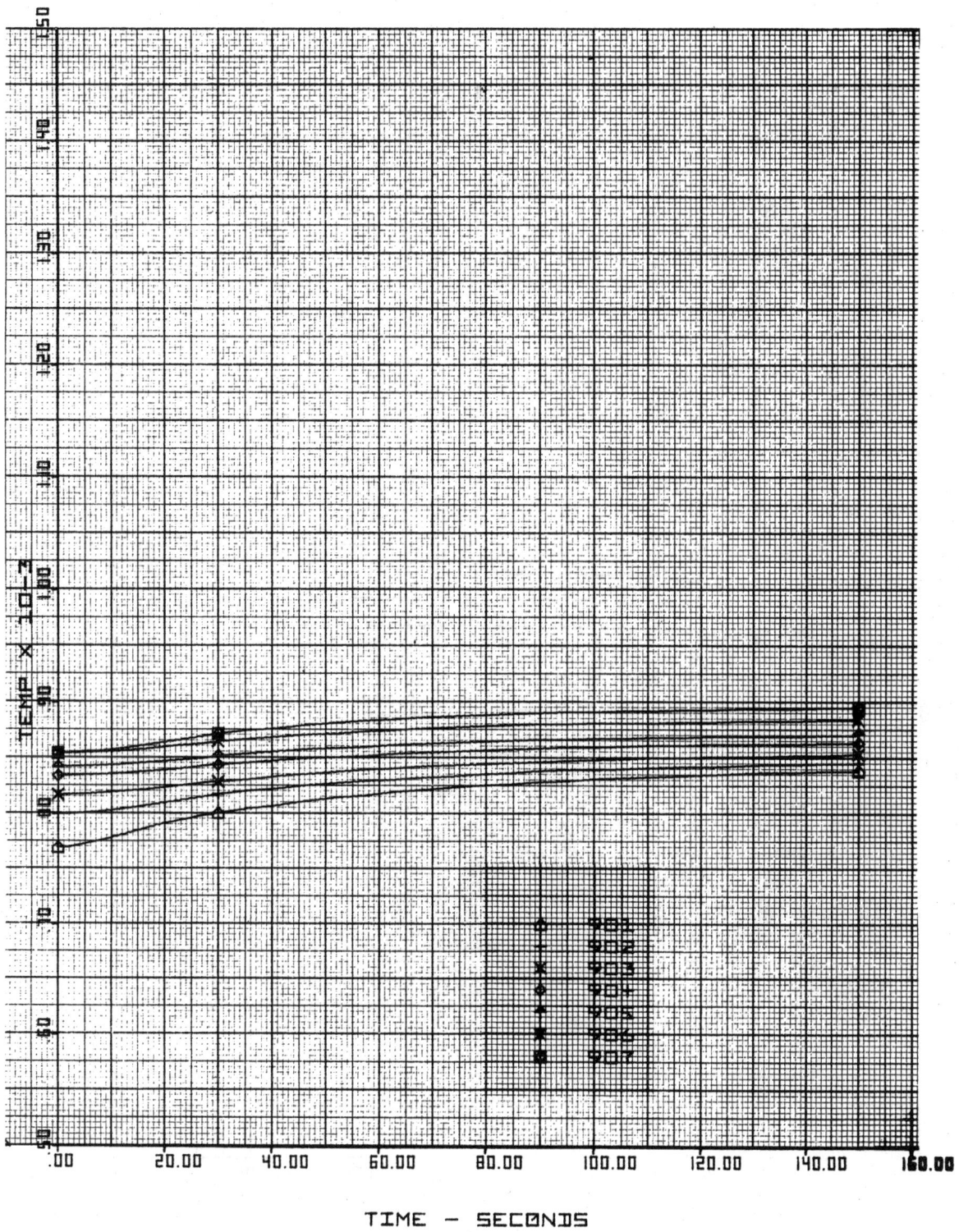

Figure A35.   Disk - Reburst.

151

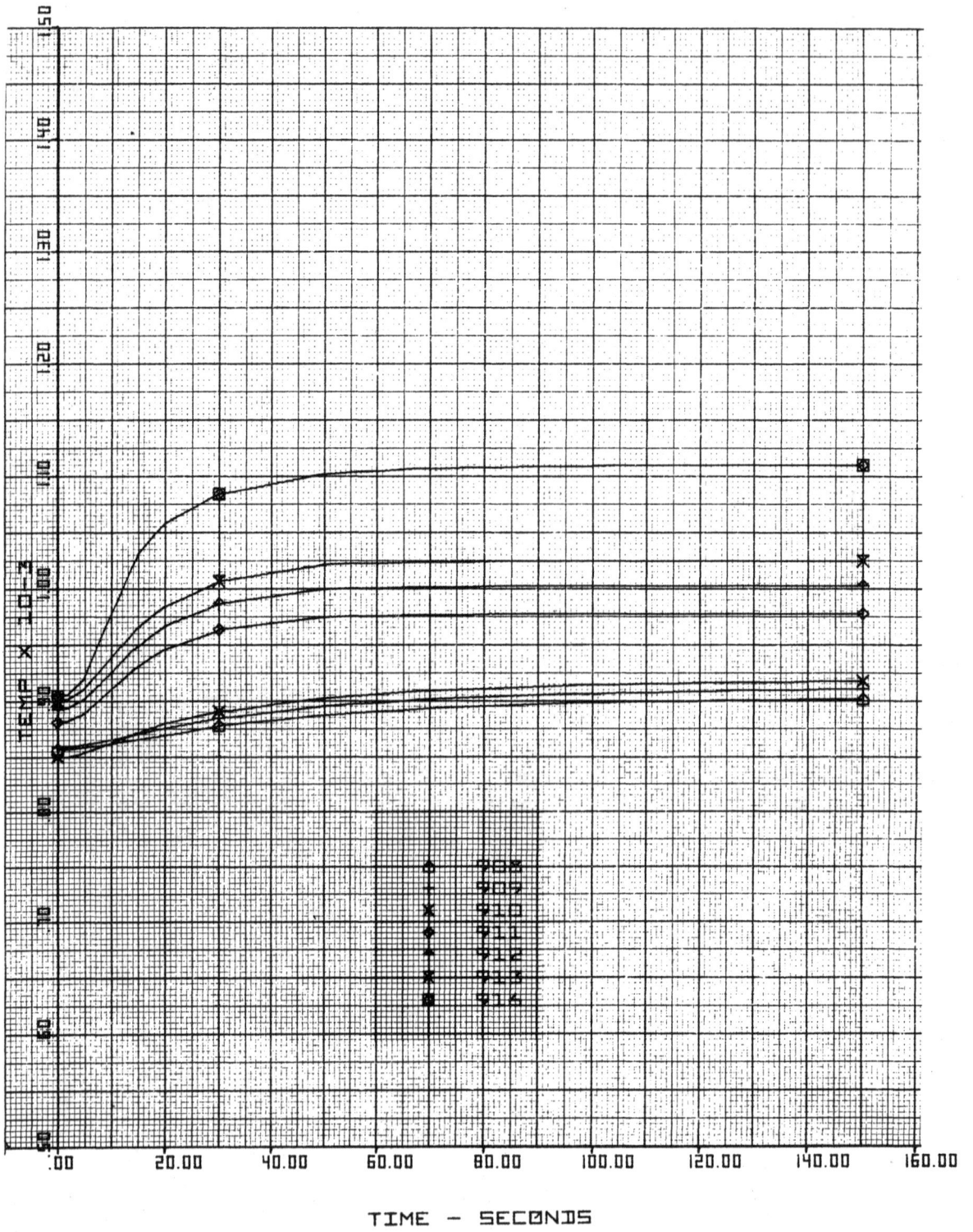

Figure A36.   Disk — Reburst.

152

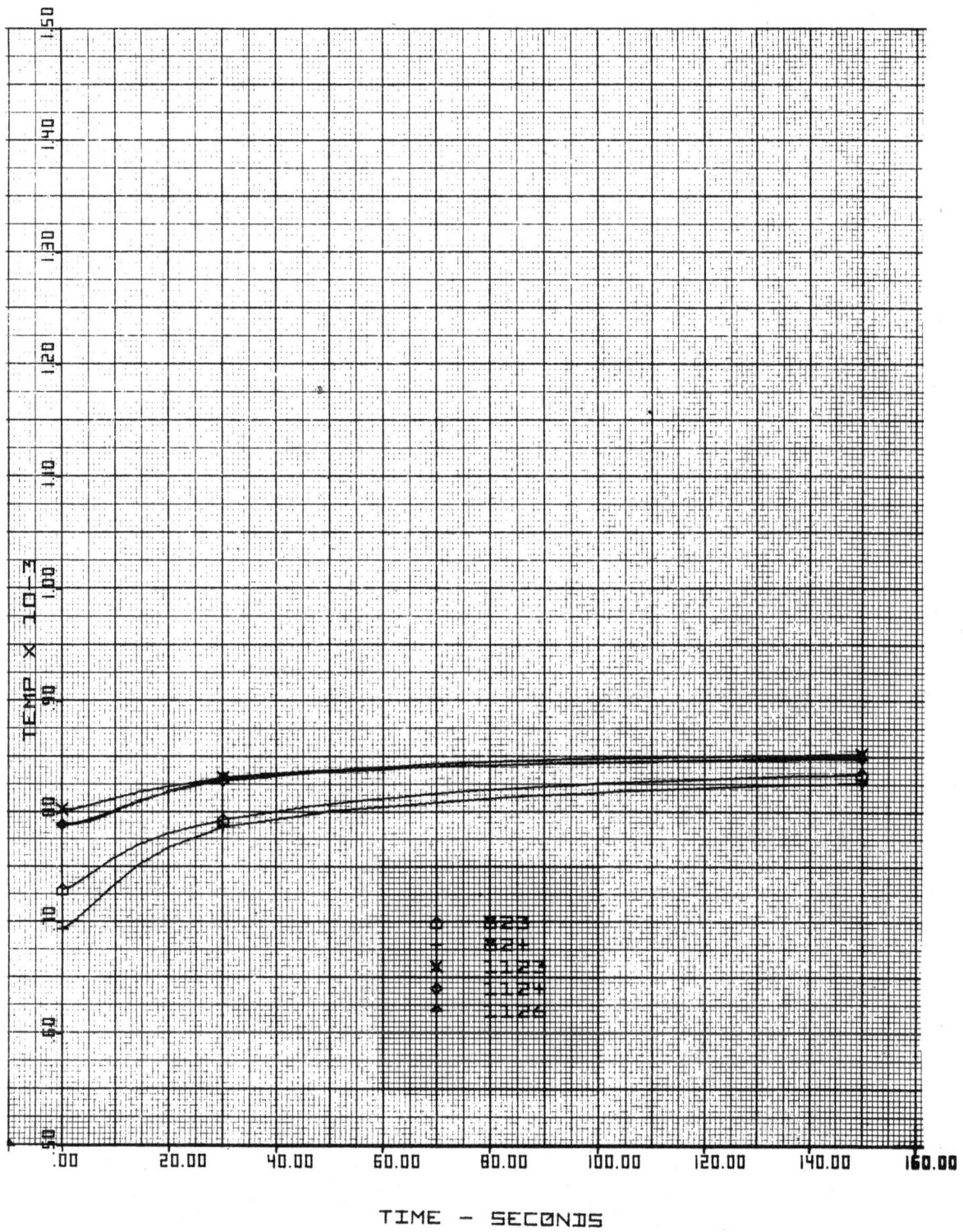

Figure A37. Miscellaneous Nodes - Reburst.

153

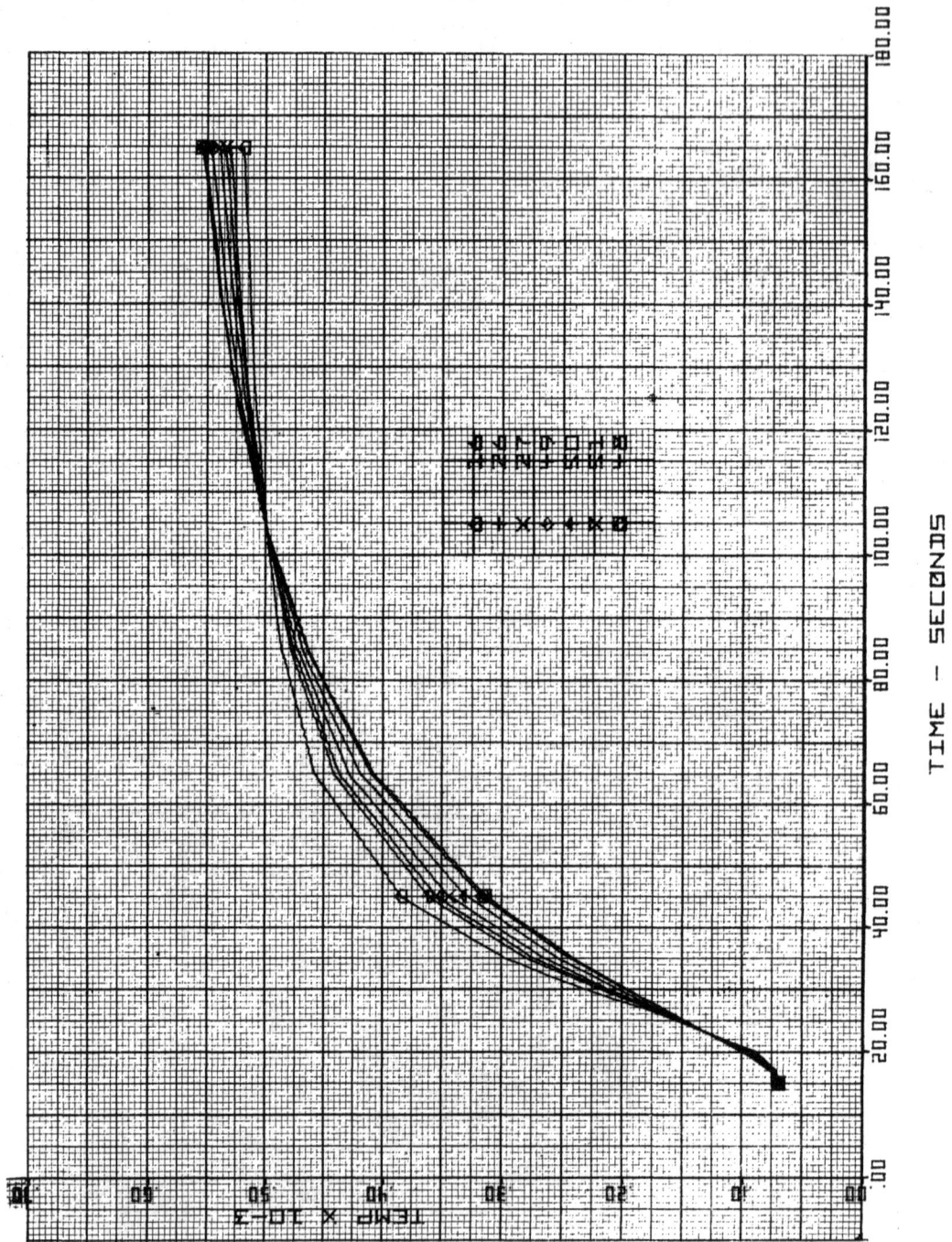

Figure A38. Stator Forward-Aft Connector - Cold Start.

154

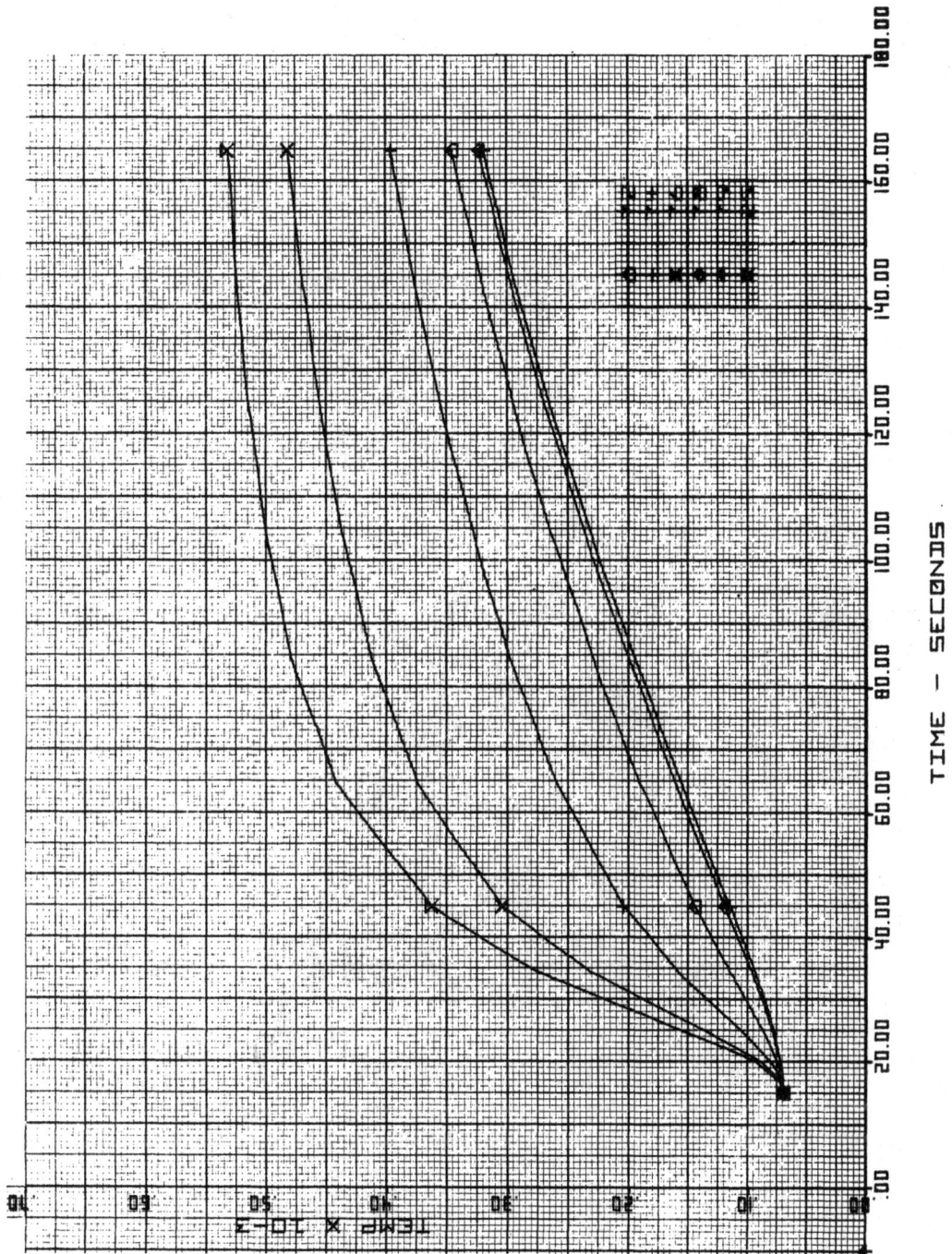

Figure A39. Stator Forward Support - Cold Start.

155

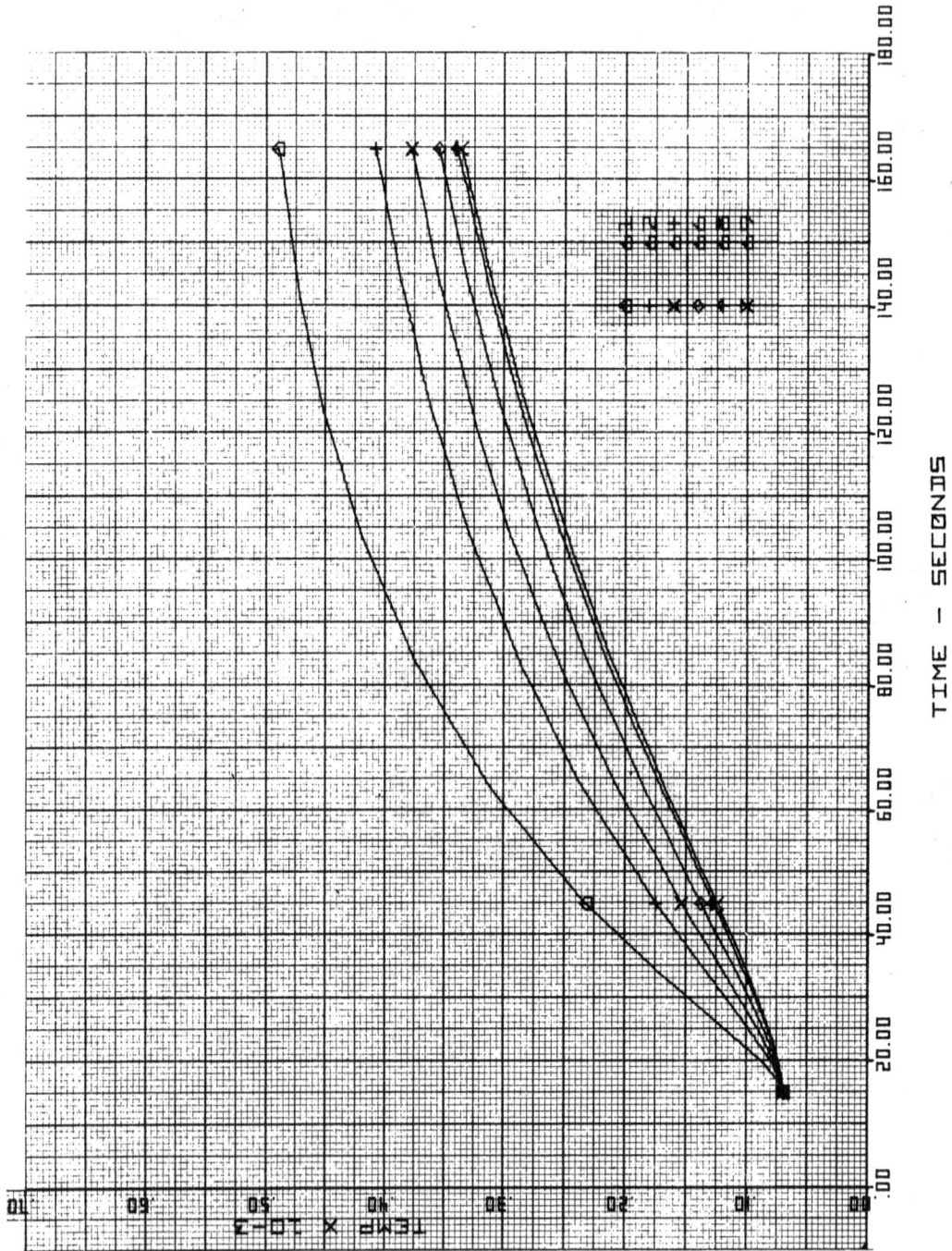

Figure A40. Stator Aft Support - Cold Start.

156

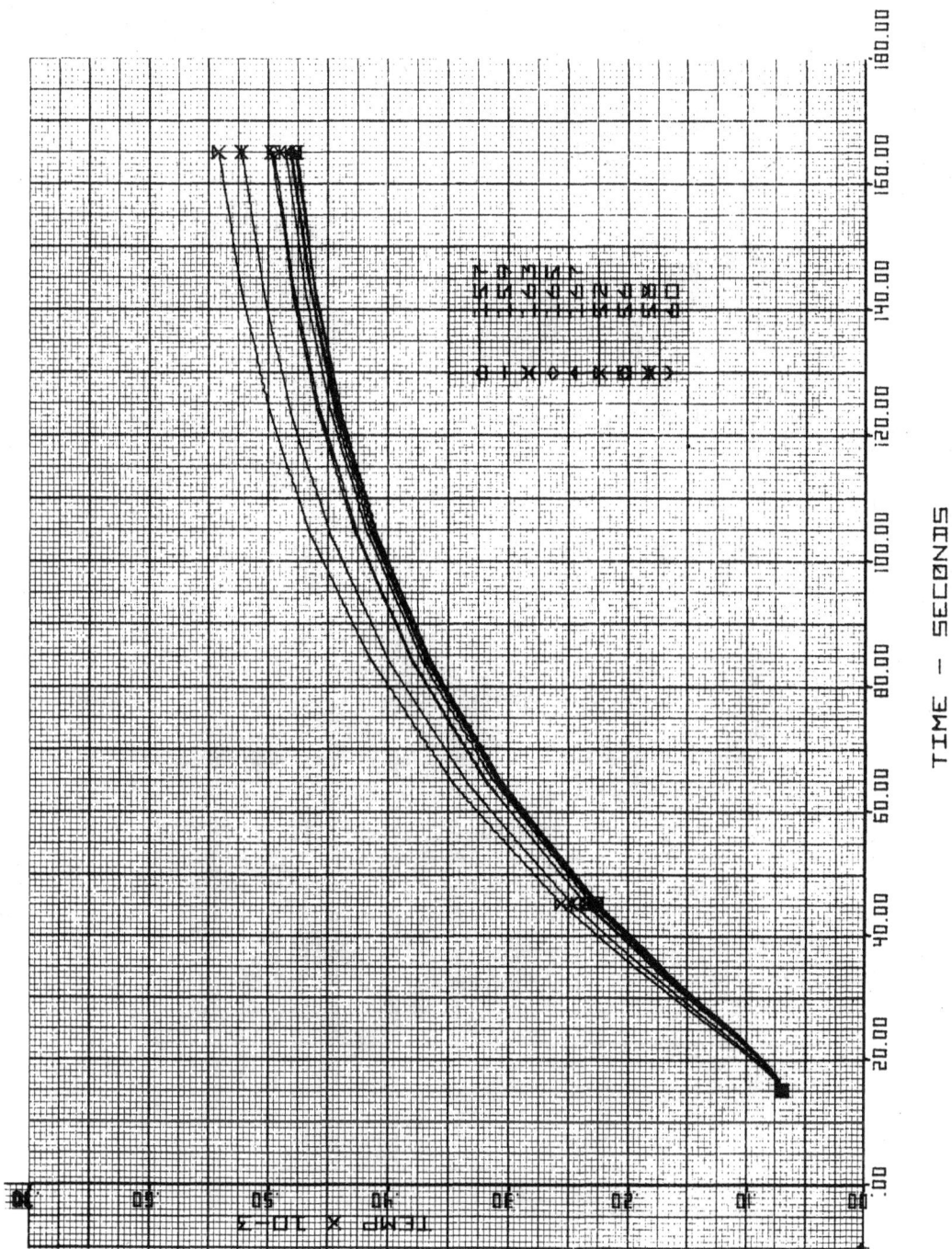

TIME – SECONDS

Figure A41. Stator Aft Support – Cold Start.

157

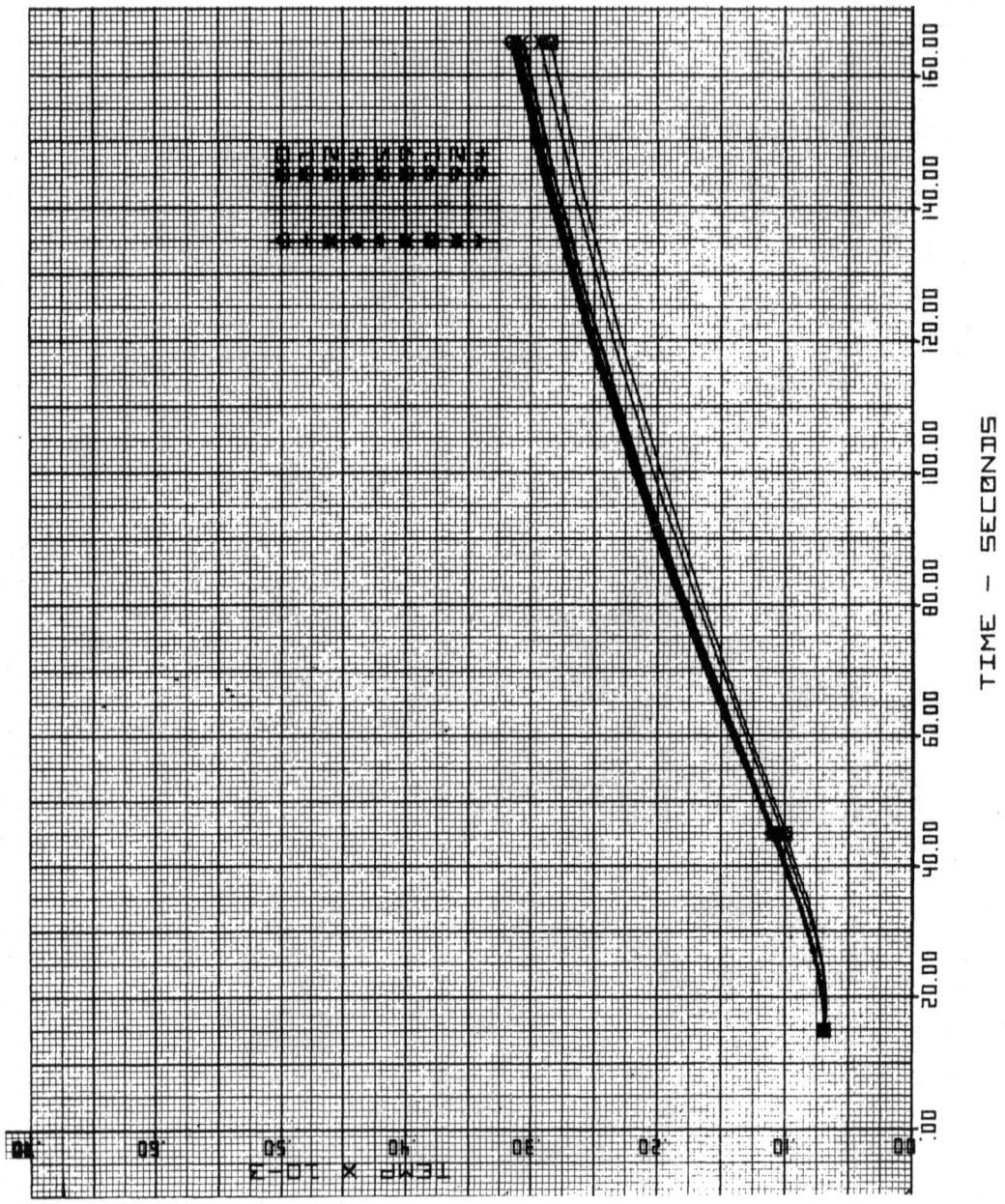

Figure A42. Stator Flange - Cold Start.

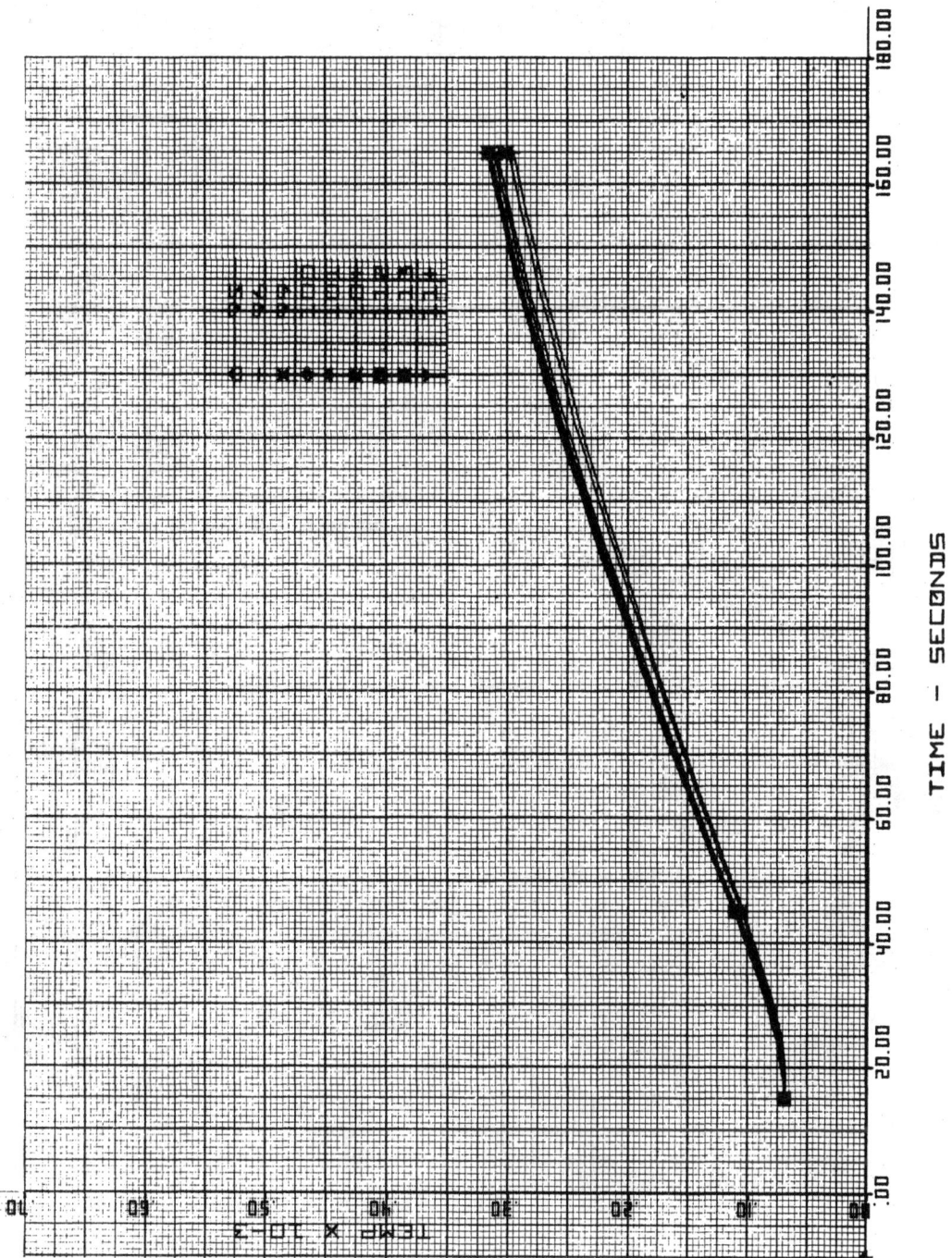

TIME - SECONDS

Figure A43. Stator Flange - Cold Start.

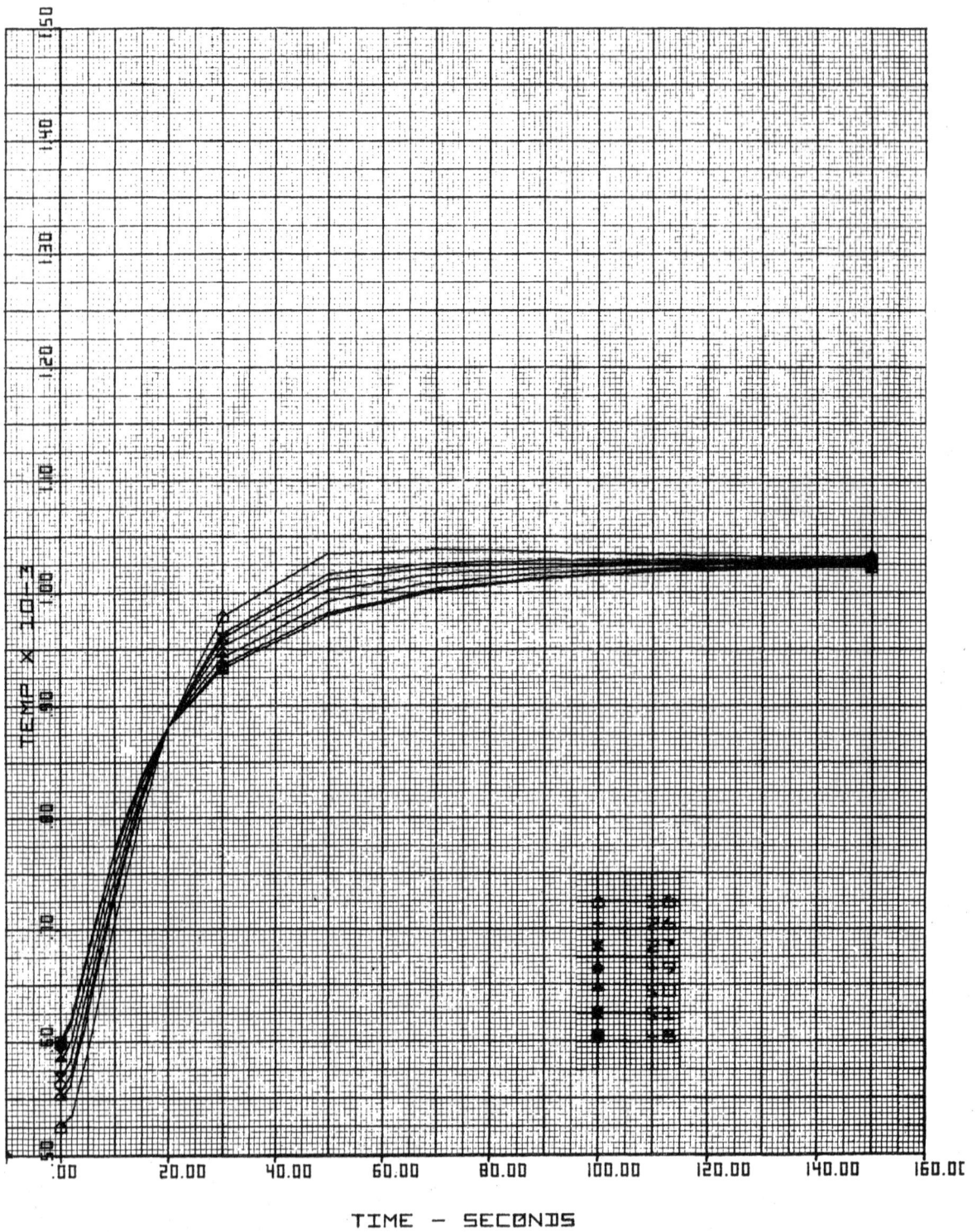

Figure A44. Stator Forward-Aft Connector - Burst.

160

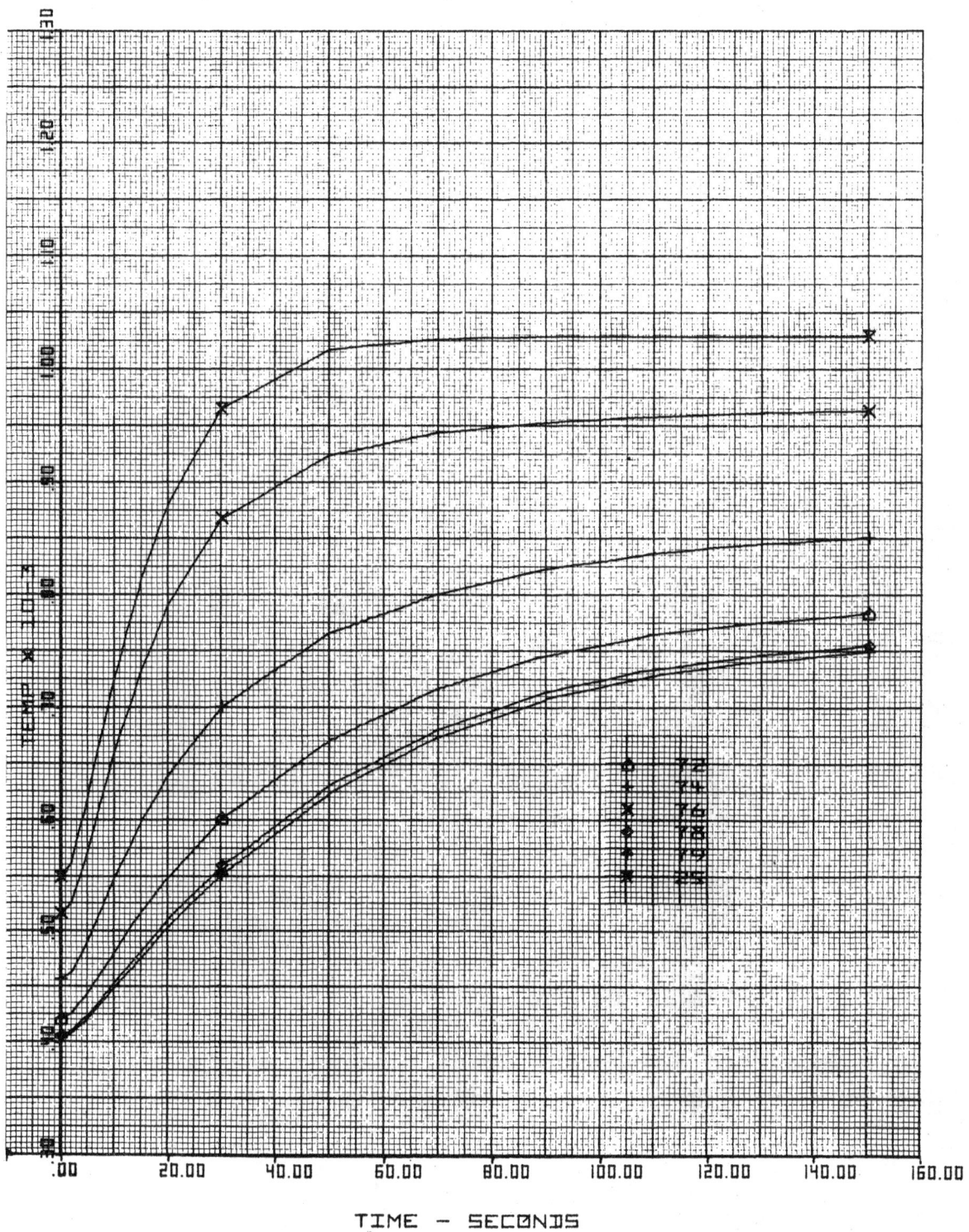

Figure A45. Stator Forward Support - Burst.

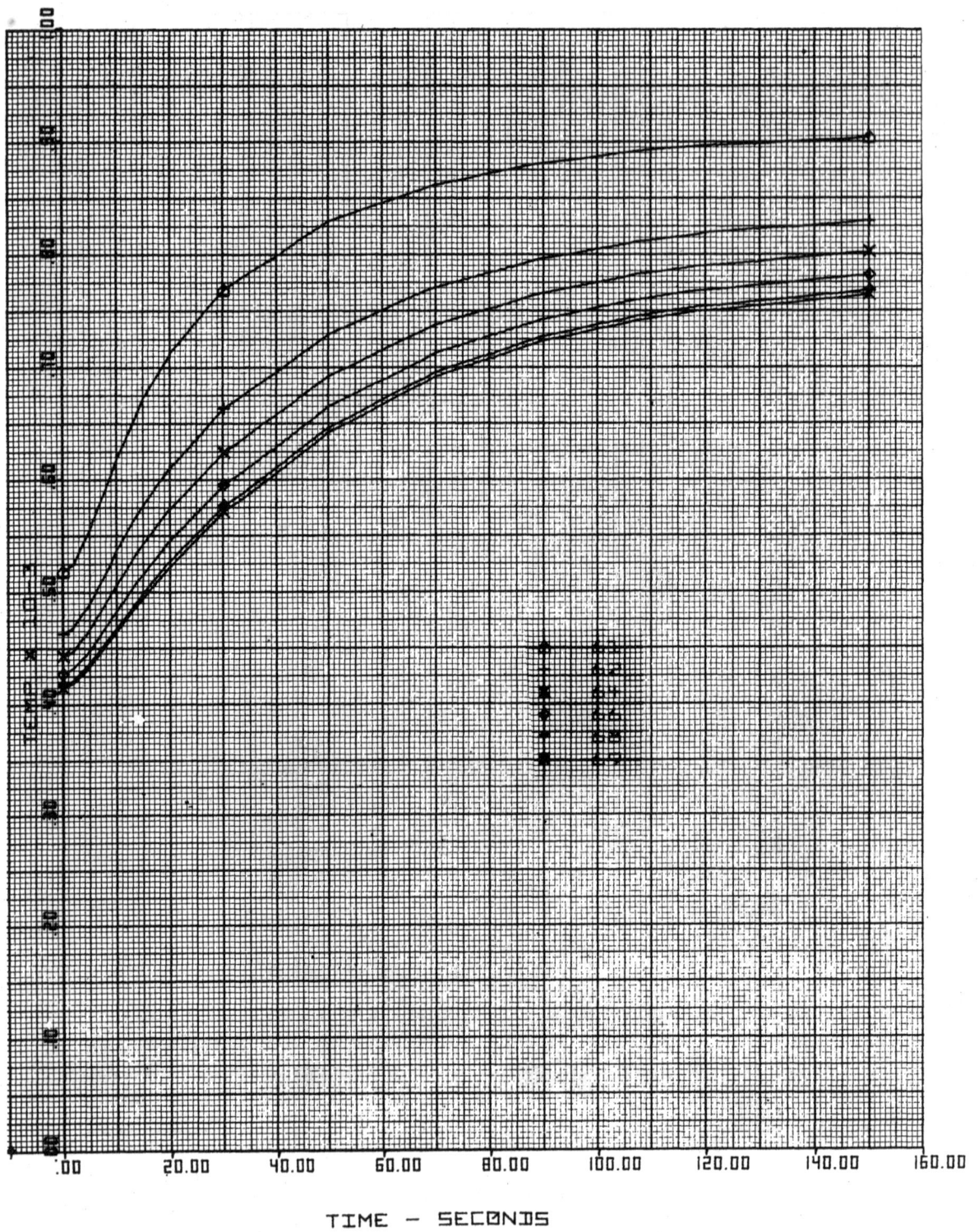

TIME — SECONDS

Figure A46   Stator Aft Support - Burst.

162

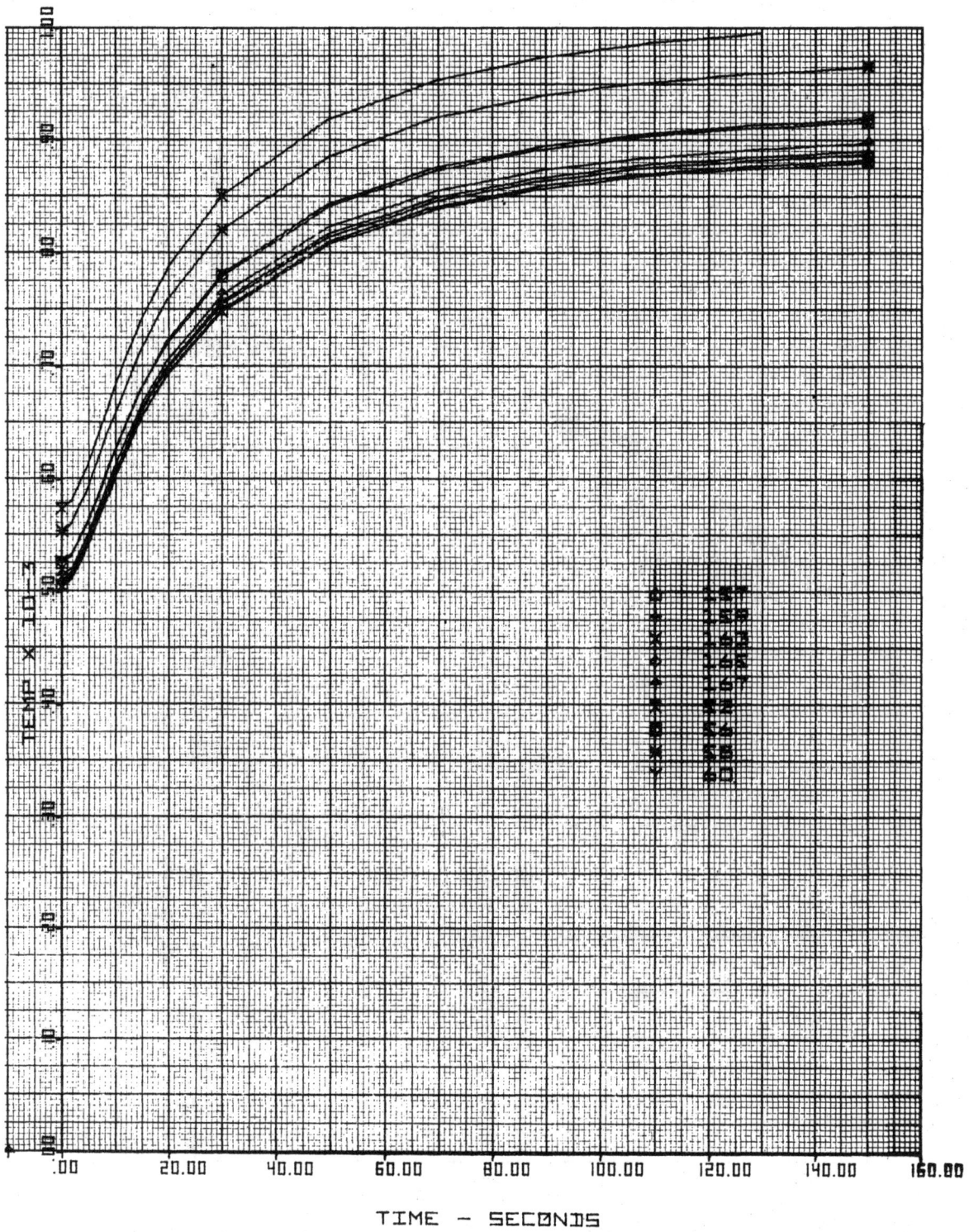

Figure A47. Stator Aft Support - Burst.

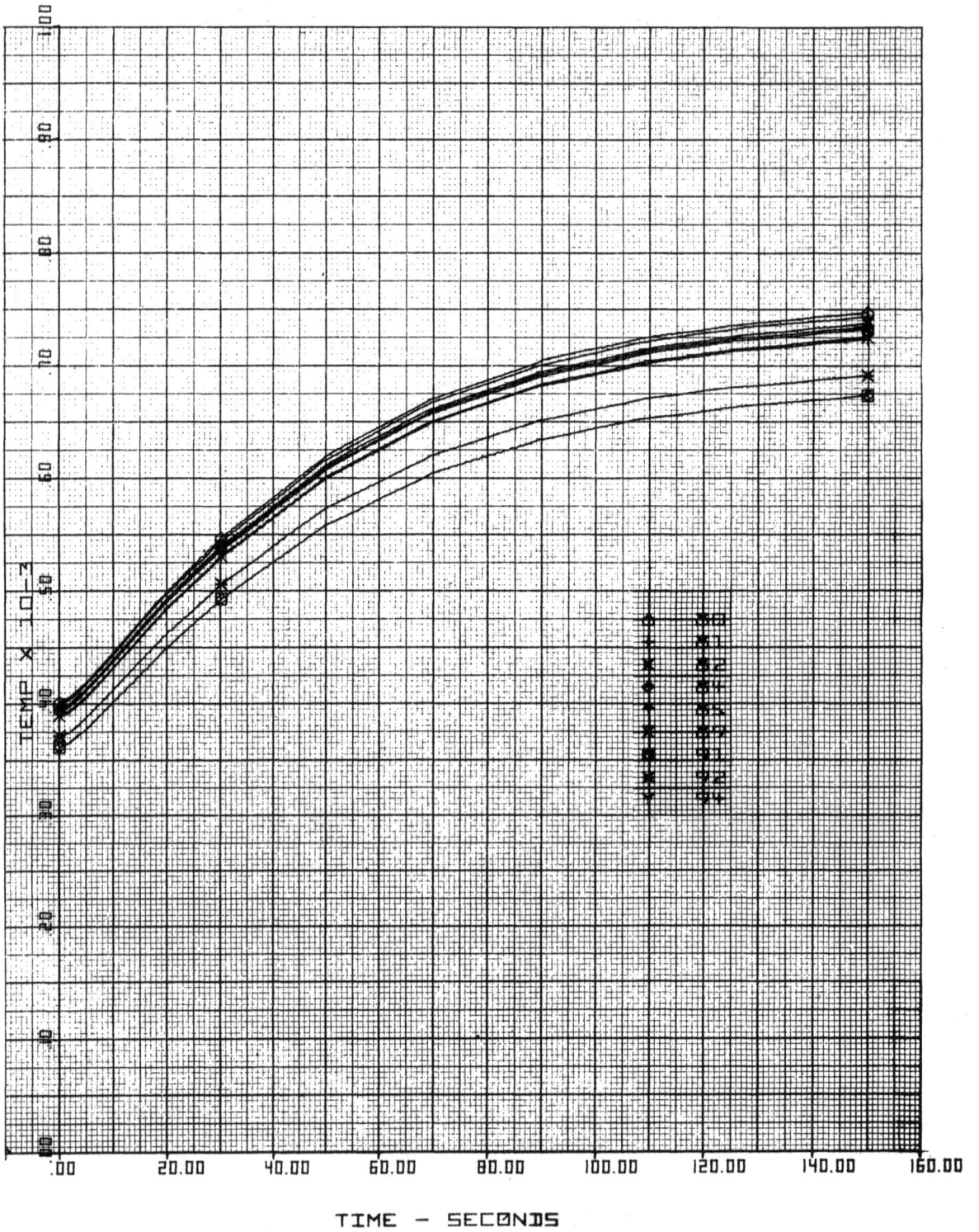

Figure A48. Stator Flange - Burst.

164

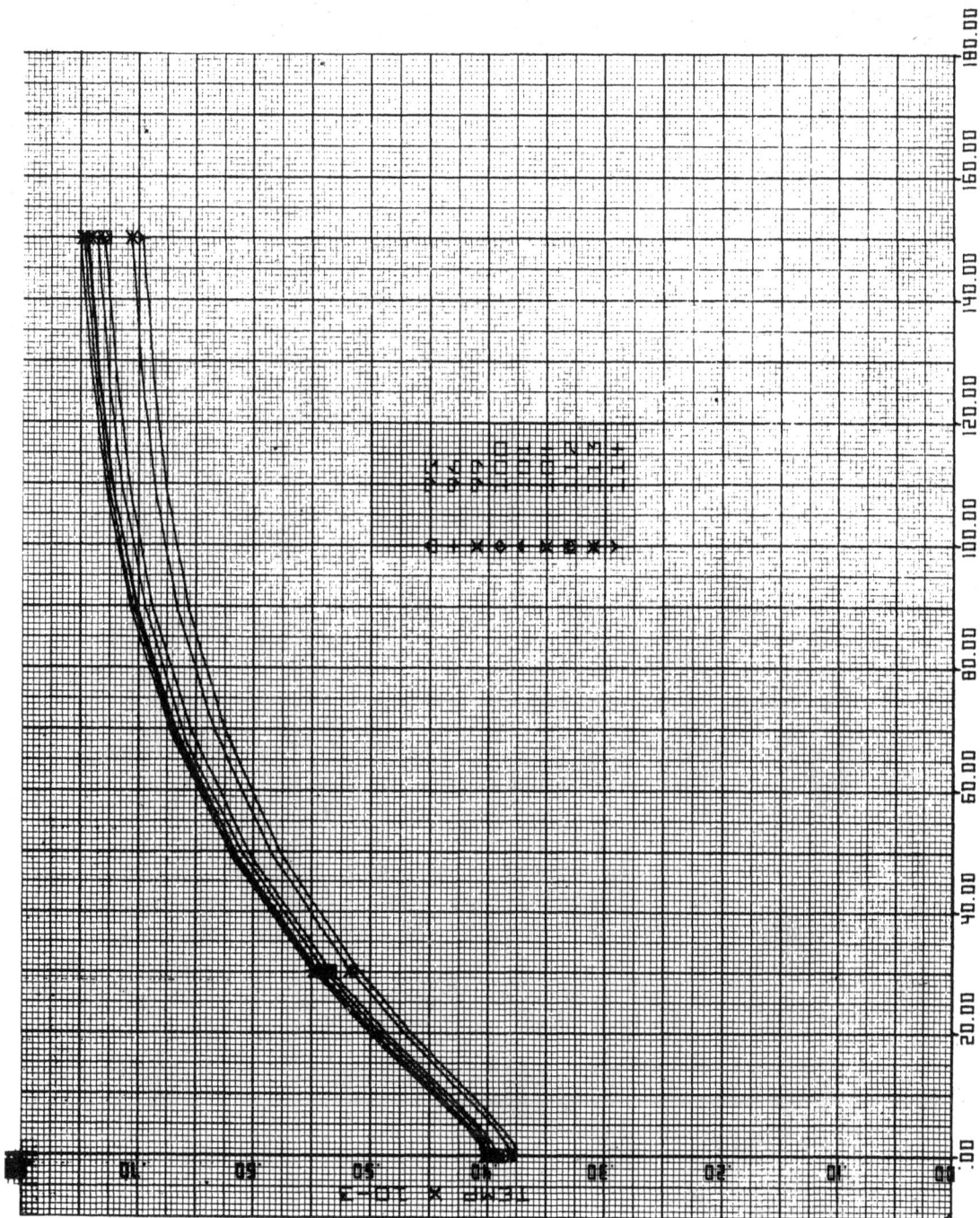

TIME — SECONDS
Figure A49. Stator Flange — Burst.

165

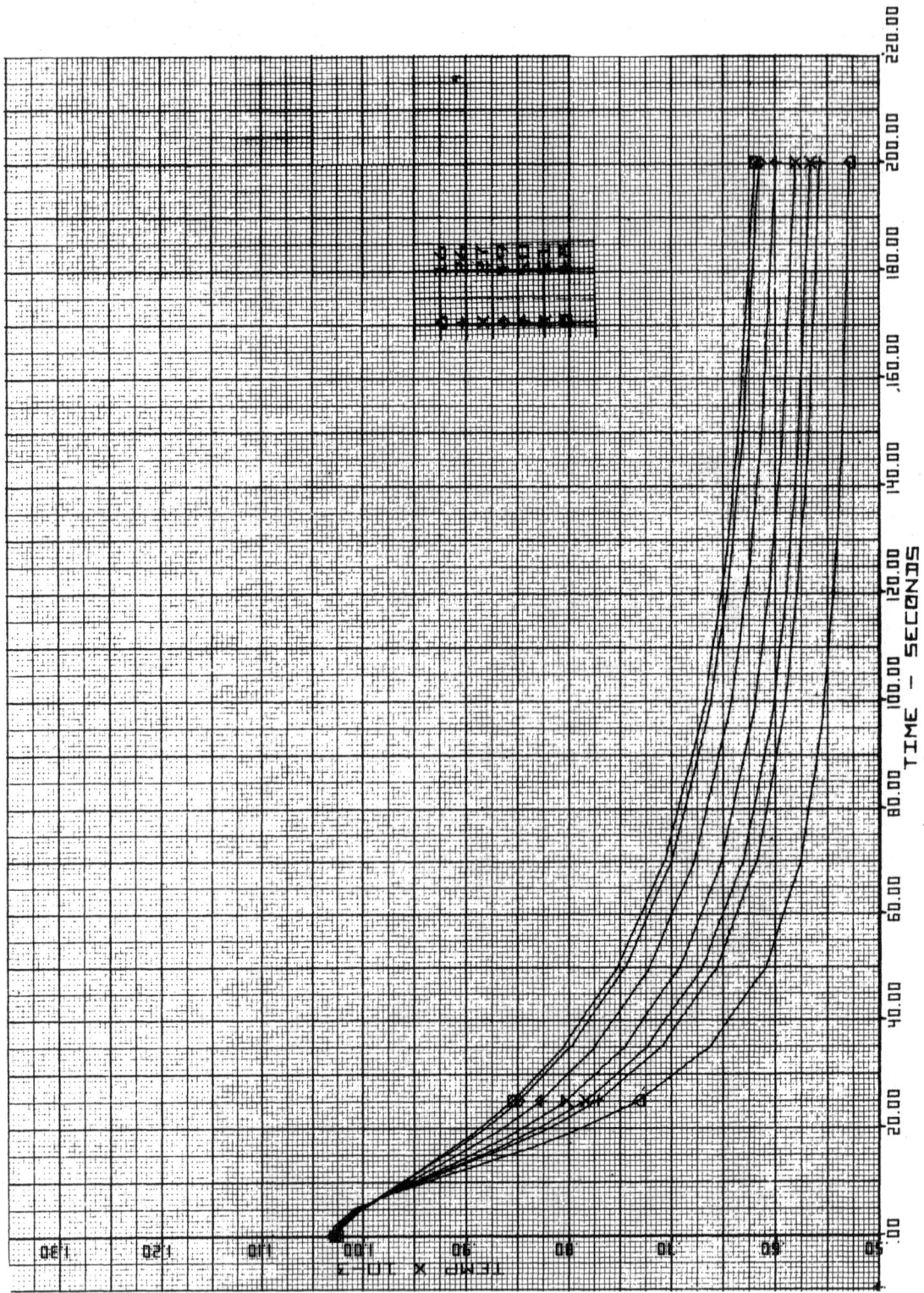

Figure A50. Stator Forward-Aft Connector - Chop.

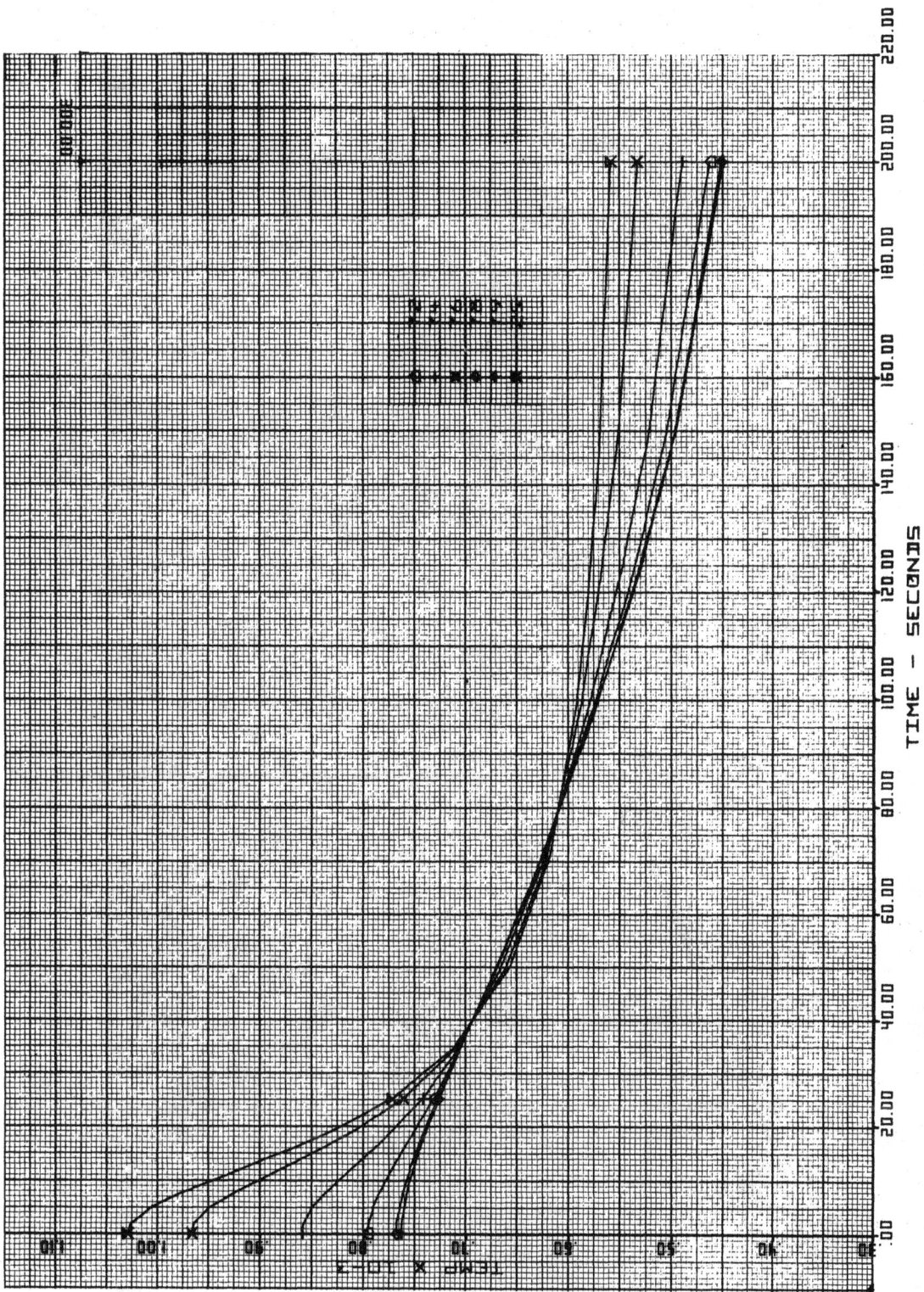

Figure A51. Stator Forward Support – Chop.

167

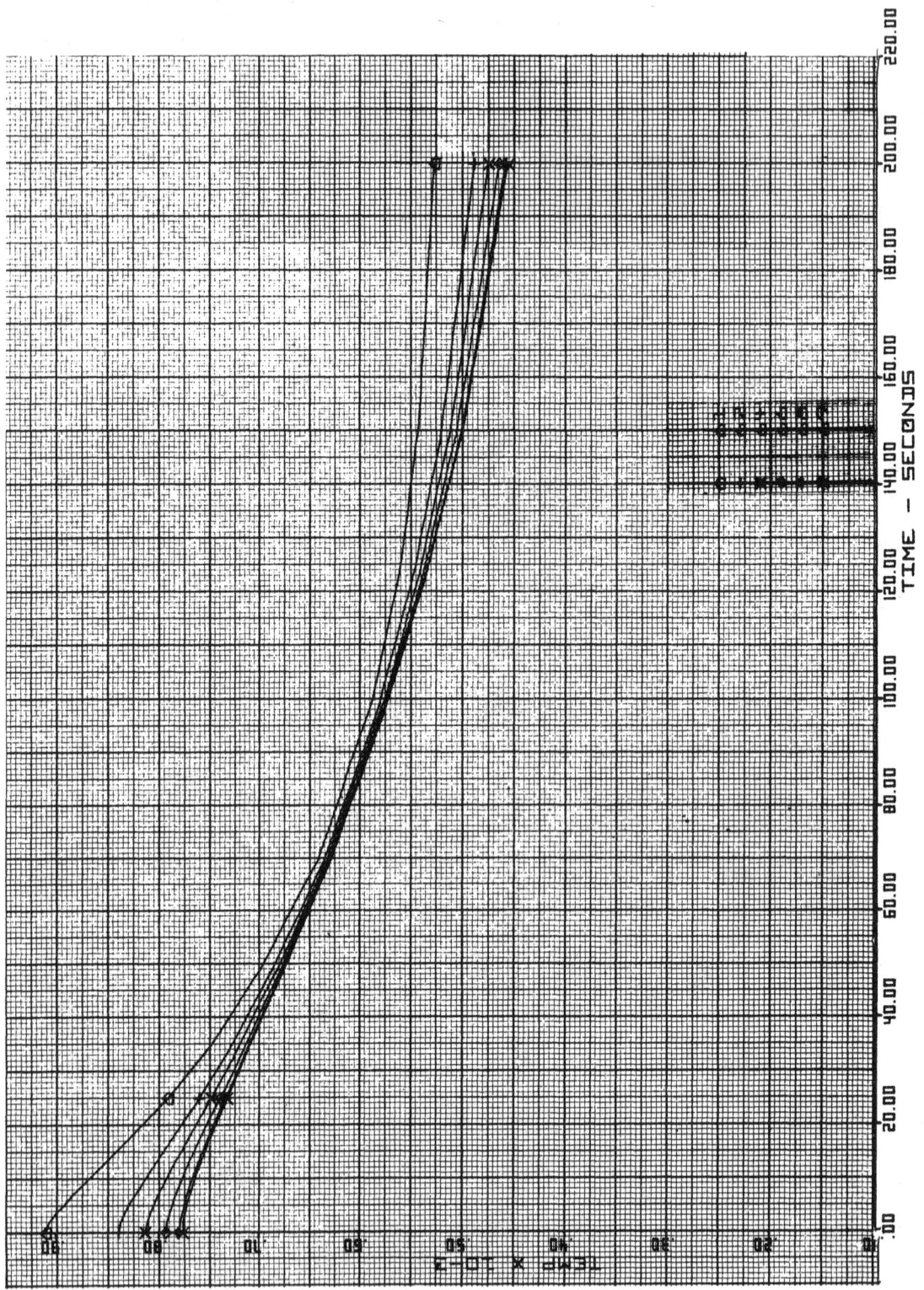

Figure A52. Stator Aft Support - Chop.

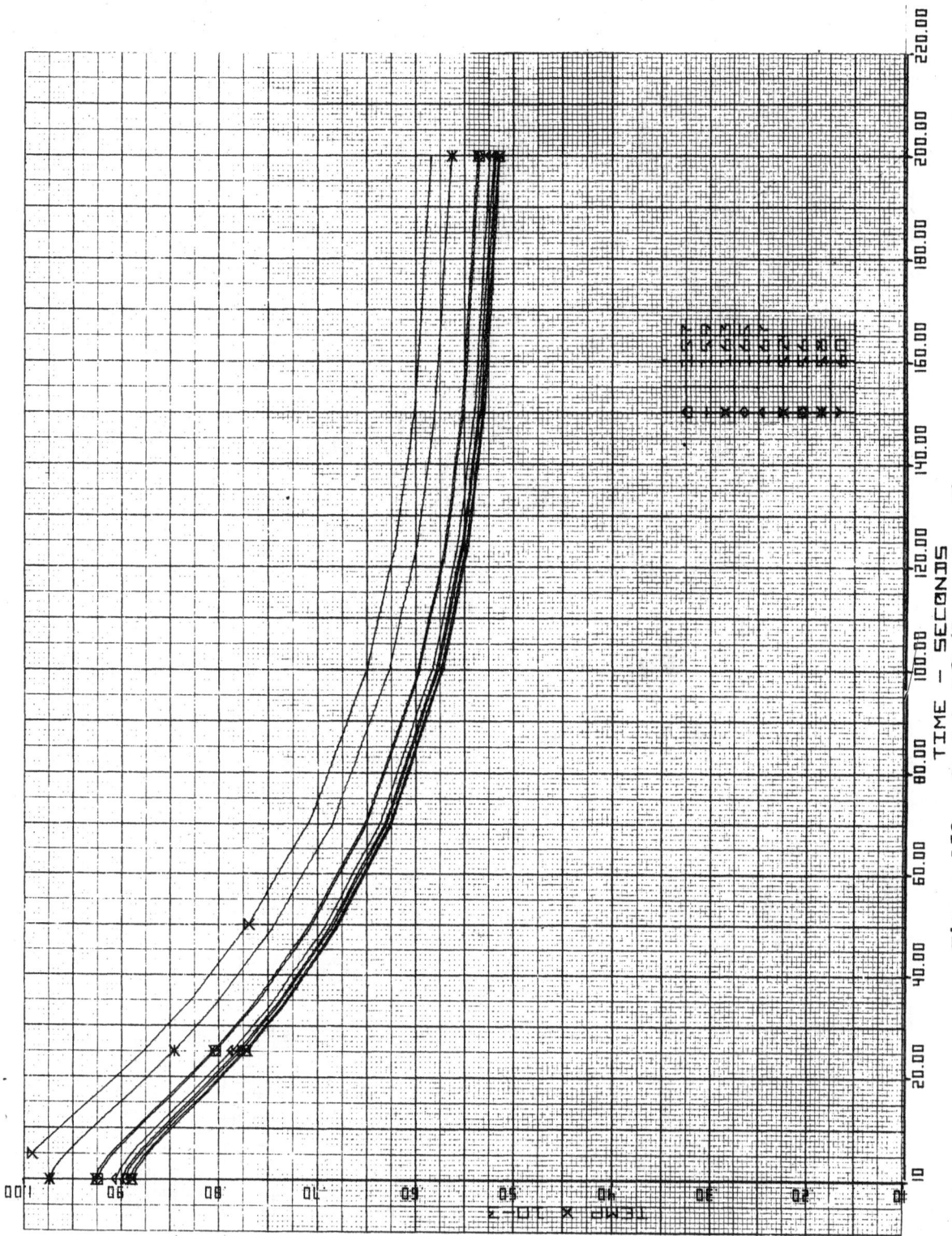

Figure A53. Stator Aft Support – Chop.

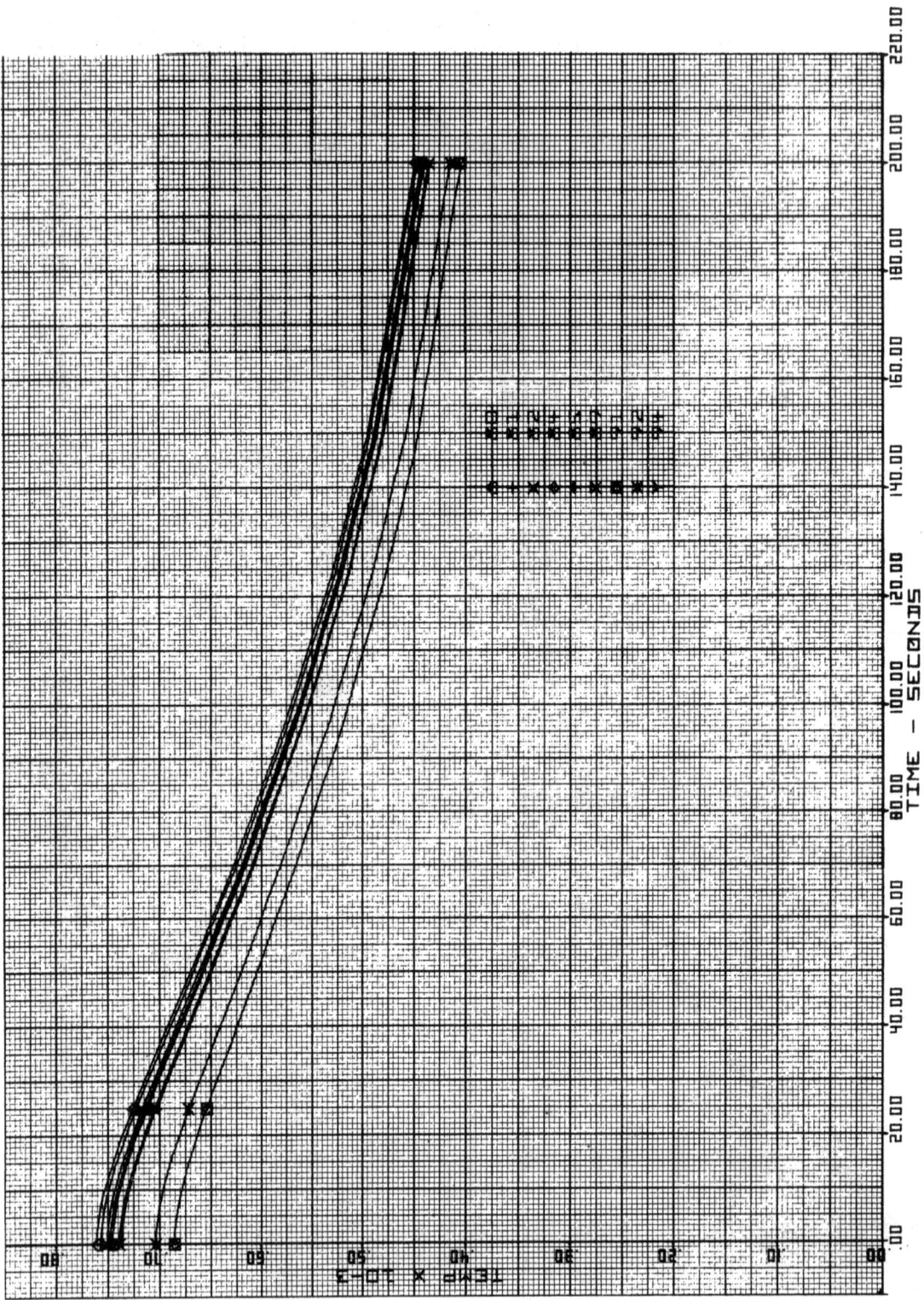

Figure A54. Stator Flange - Chop.

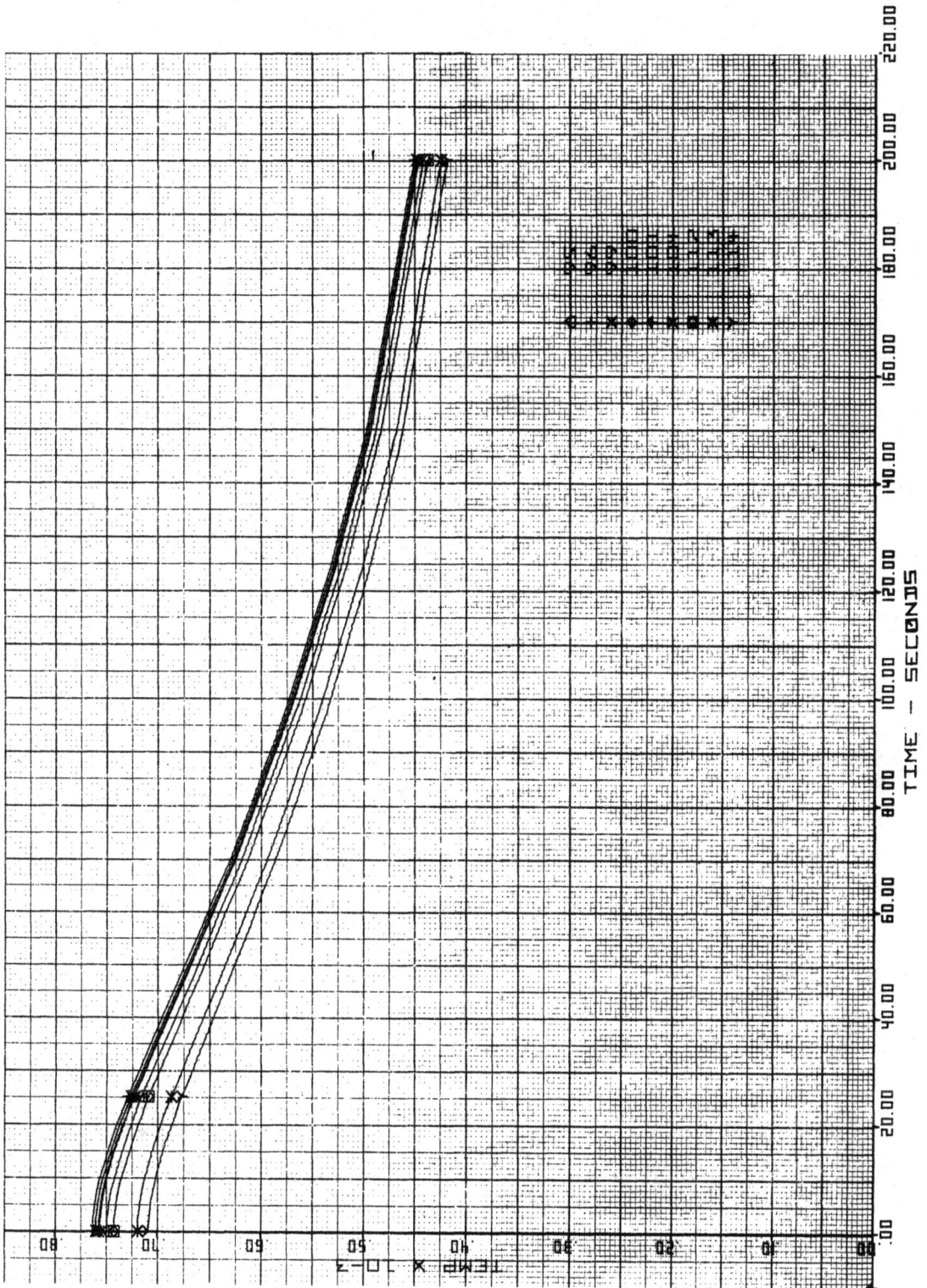

Figure A55. Stator Flange – Chop.

171

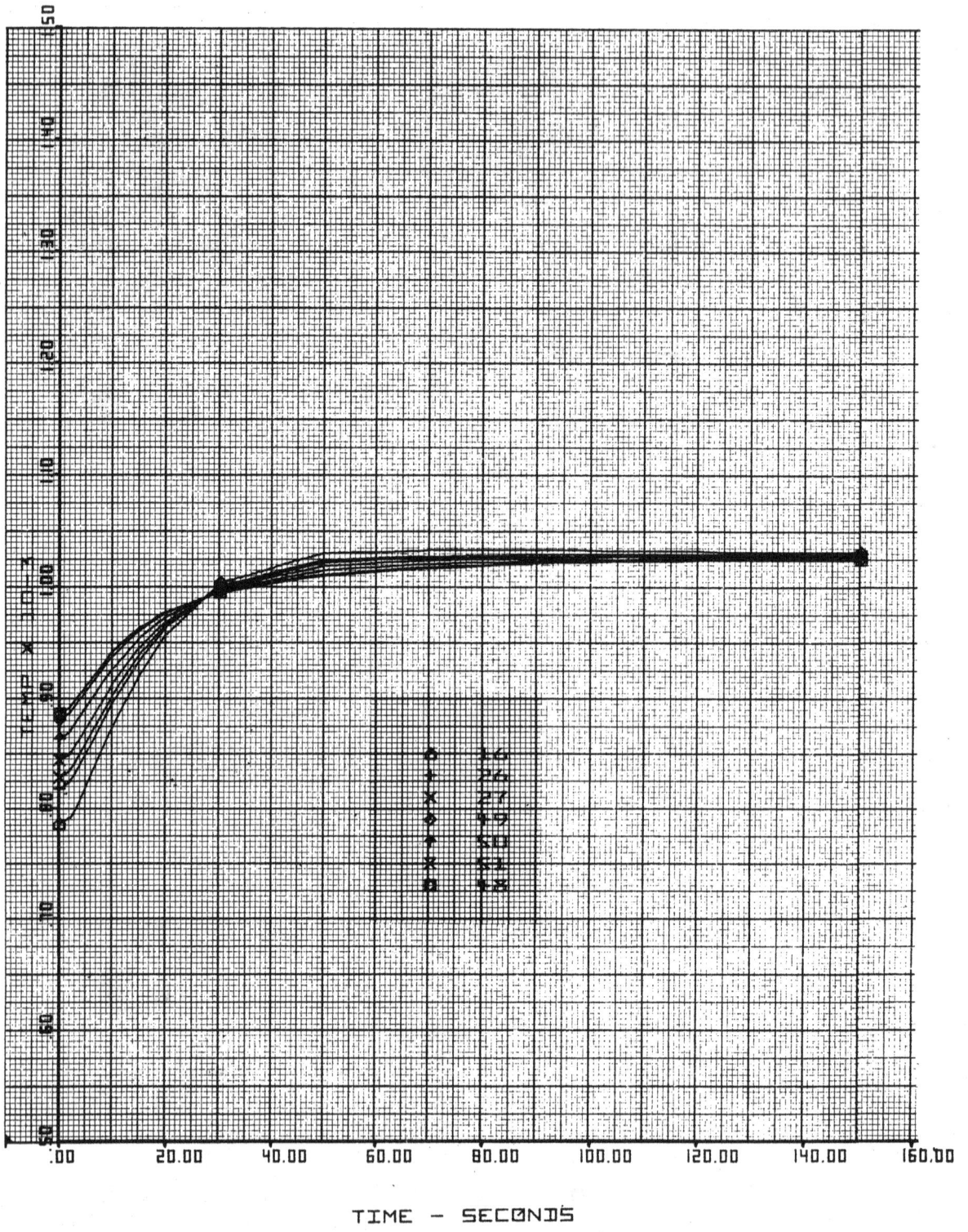

TIME — SECONDS

Figure A56. Stator Forward-Aft Connector - Reburst.

172

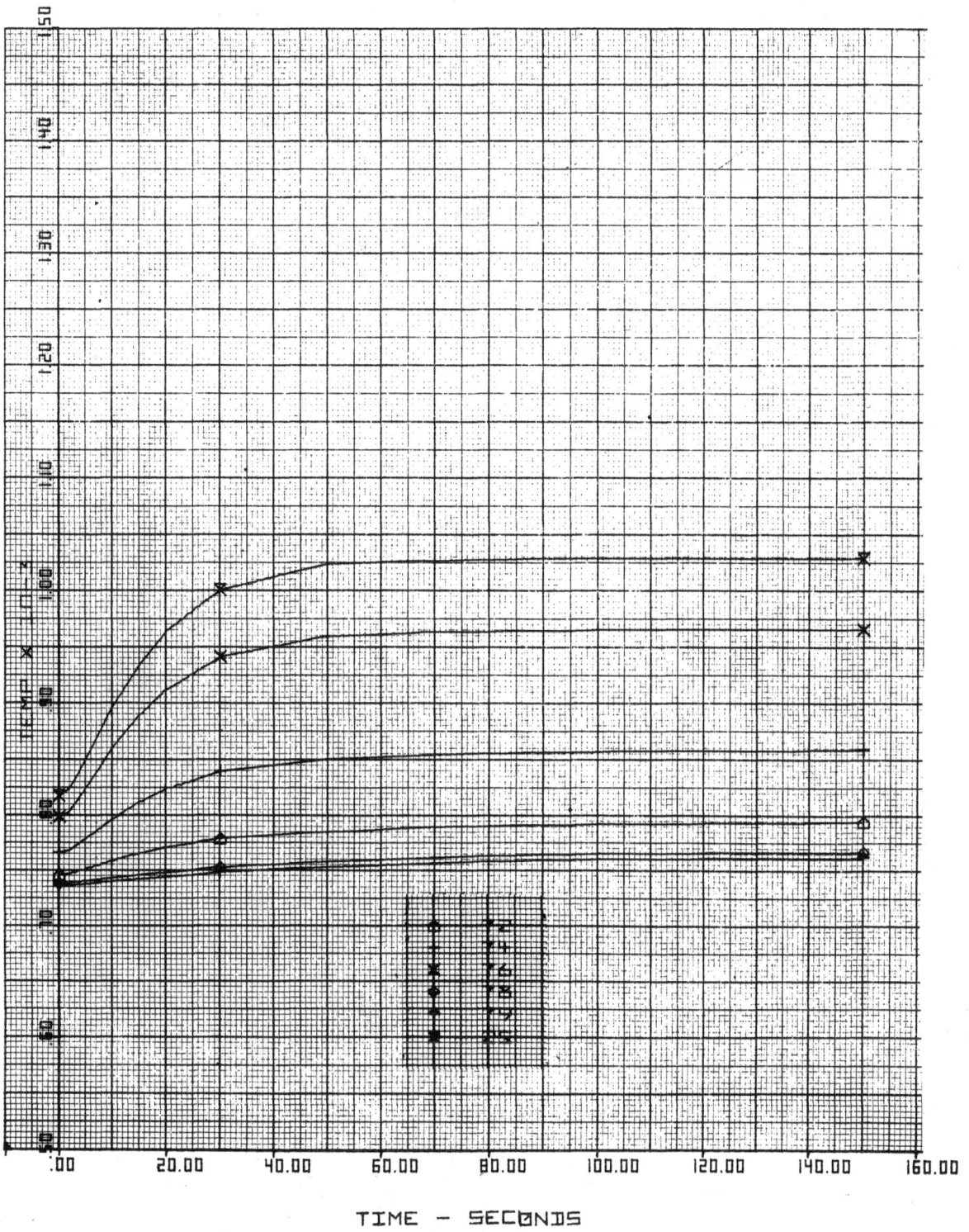

Figure A57. Stator Forward Support - Reburst.

173

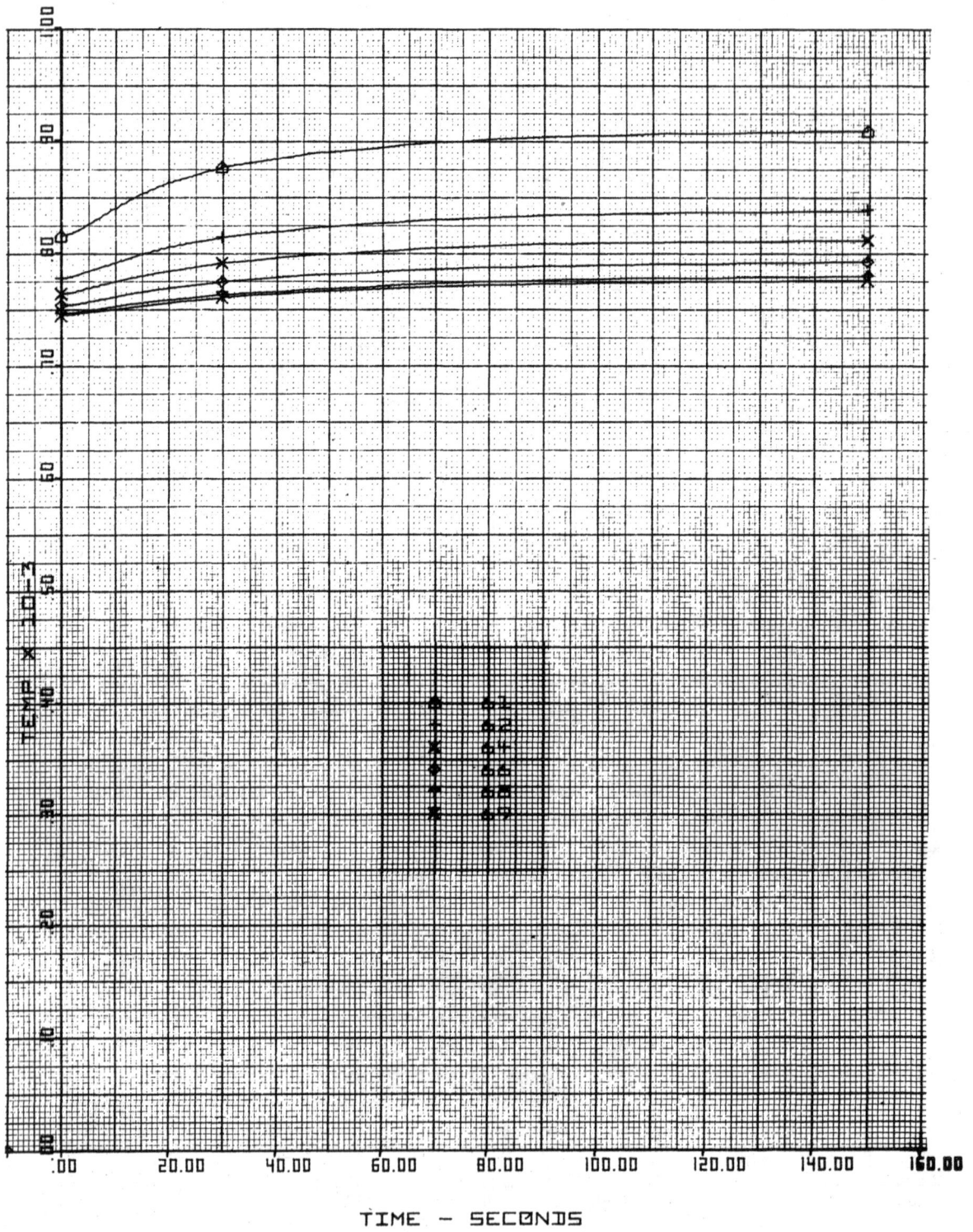

Figure A58. Stator Aft Support - Reburst.

174

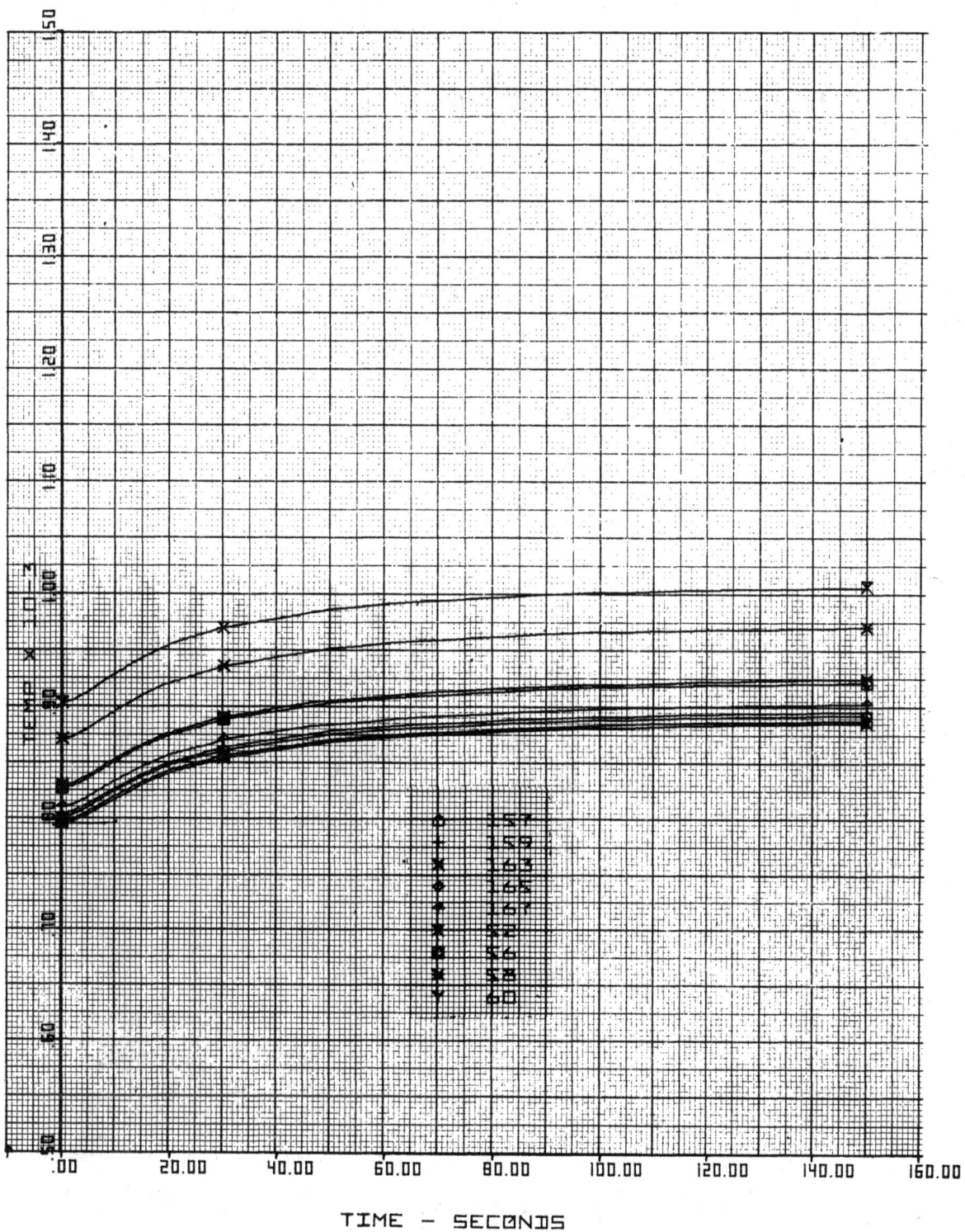

Figure A59.   Stator Aft Support - Reburst.

175

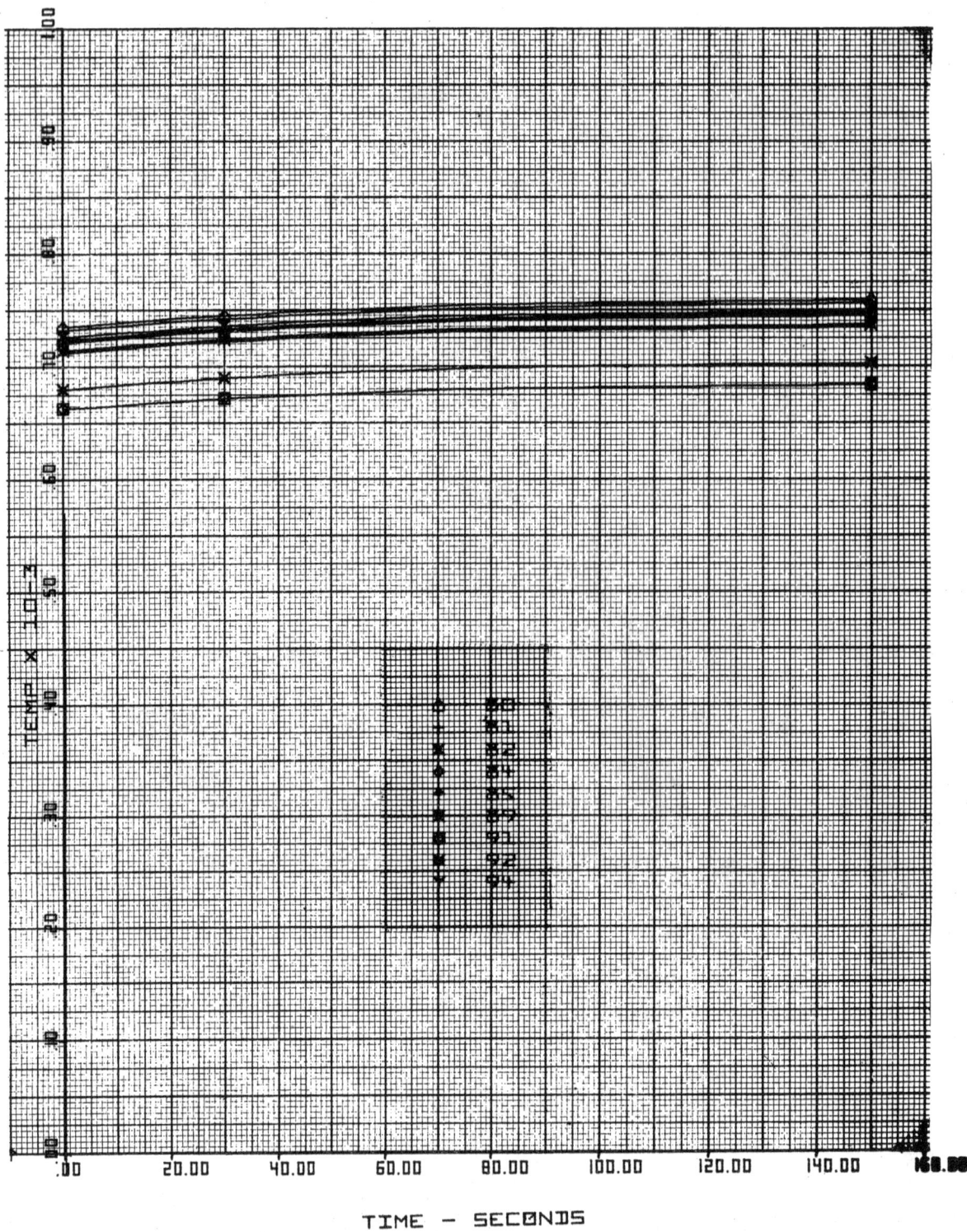

Figure A60. Stator Flange - Reburst.

176

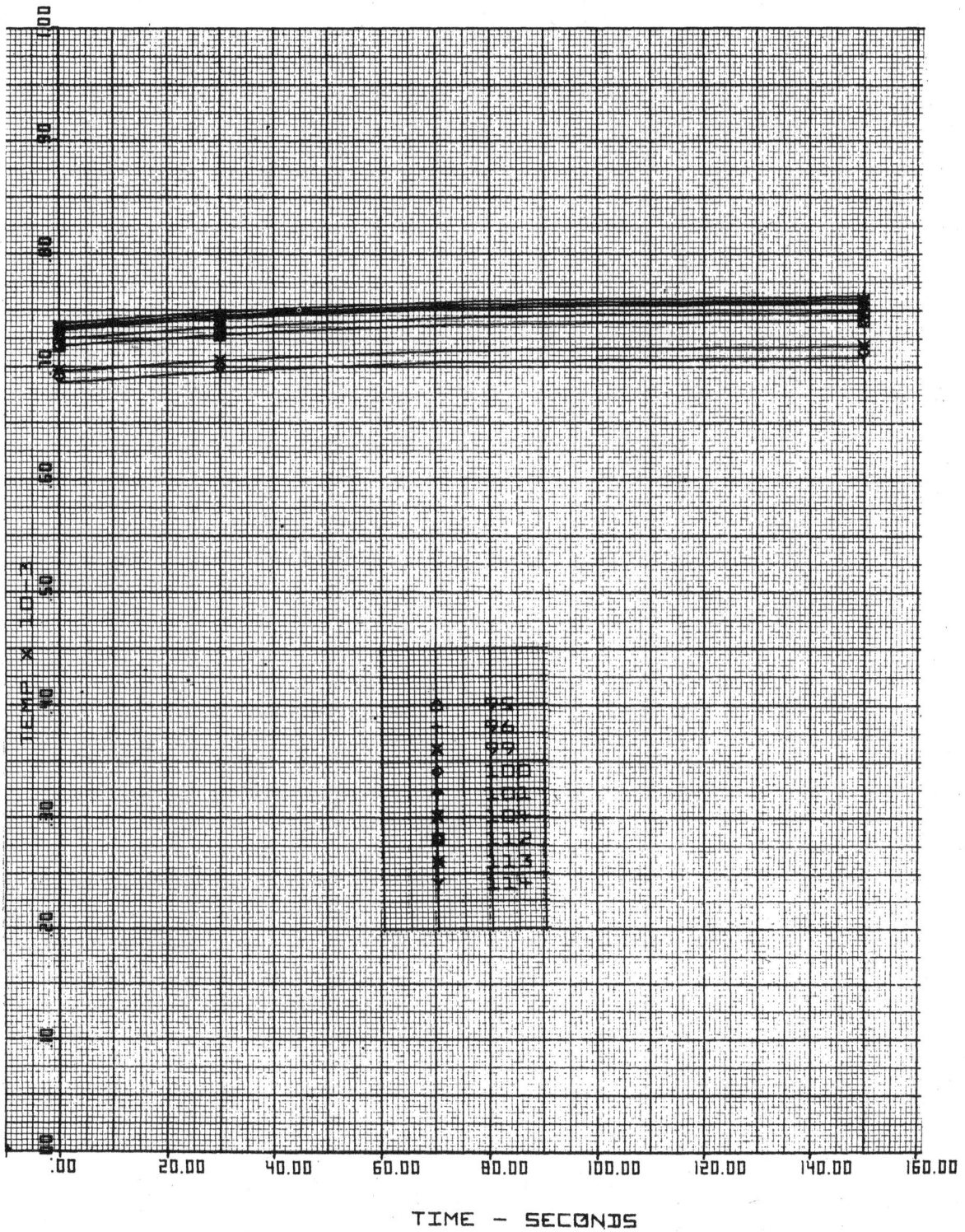

Figure A61.　Stator Flange - Reburst.

3896-78